The Third Option for the South China Sea

David Jay Green

The Third Option for the South China Sea

The Political Economy of Regional Conflict and Cooperation

palgrave
macmillan

David Jay Green
Hult International Business School
San Francisco, California United States

ISBN 978-3-319-40273-4 ISBN 978-3-319-40274-1 (eBook)
DOI 10.1007/978-3-319-40274-1

Library of Congress Control Number: 2016947459

Printed on acid-free paper

This Palgrave Macmillan imprint is published by Springer Nature
The registered company is Springer International Publishing AG Switzerland

For my children, Naomi and Lucian Wizer-Green, who grew to young adulthood on the shores of the South China Sea

Foreword

David Green and I were colleagues at the Asian Development Bank in Manila for 20 years (full disclosure). He has always been bitten by the curiosity bug, and has had an admirable instinct for seeking intelligent responses to vexed questions—traits not typically associated with formulaic multilateral development finance institutions. This inquiry into the prospects for regional cooperation in Southeast Asia is unsurprising—David knows the issues well, and he has framed them in the context of the complex political and security overtones that have developed in recent times.

It is not easy to add value to a subject as fraught as the goings-on in the South China Sea. Events leapfrog each other with surprising rapidity, fig leafs are off, militarization proceeds apace, and the room for political maneuver narrows. It is hard to predict what happens next, but it is likely that the asymmetry of interests, political strength, and negotiating abilities will assure mutual self-preservation. Temperatures in the cauldron will rise, but open conflict will probably be avoided.

David's thesis acquires appeal in this background, and merits a closer look. The benefits of regional economic cooperation, particularly in a Factory Asia environment, are unique and independent. It focuses attention on economic growth and social development. Participating countries tend to subsume often narrowly conceived national interests for the greater good. Short-term gains are occasionally traded off for long-term ones. A sense of political maturity develops over time as a regional identity begins to emerge. Certainly, this has been the experience in Southeast Asia, and David has witnessed it from close quarters.

Regional economic cooperation in Southeast Asia is not new. The Asian Development Bank pioneered the Greater Mekong Subregion Economic Cooperation Program in 1992. About 25 years later and with billions of dollars in investments, regional infrastructure has taken shape, cross-border trade has grown manyfold, and cooperation in sectors such as agriculture, health, and tourism has taken off. Other programs such as the Indonesia-Malaysia-Thailand Growth Triangle, or the Brunei-Indonesia-Malaysia-Philippines East ASEAN Growth Area, are of similar vintage but have produced less dramatic results for reasons David explains. But, clearly, conflict has been substituted by cooperation and shared economic and social benefits.

We then ask whether such cooperation can be infinitely successful regardless of support by multilateral development institutions. It can, but from different perspectives. The growth of Factory Asia has intensified the imperatives of economic cooperation. If Laos has hydropower resources that are of little use for itself given its low population numbers and fledgling economy, the logic of China, Thailand, and Viet Nam investing in them to secure stable power supplies makes sense in an environment where manufacturing and logistic supply chains need to be well oiled, and well fed, across the subregion. A common argument that economic cooperation in the subregion is likely to be affected by a slowing Chinese economy, especially in the context of Factory Asia, has only limited validity. Rising growth rates in the Philippines, Indonesia, and Viet Nam will ensure that the intensity of cooperation does not diminish, even if lower numbers prevail.

Perspectives notwithstanding, the pursuit of the holy grail of regional economic cooperation can hardly be claimed as a panacea to what is obviously a tangled political web. Political positions, developed on strategic considerations and national sentiment, are unlikely to be given up unilaterally anytime soon; if anything, they are likely to harden. But governments' hanging their ambitious economic growth agendas on the peg of regional cooperation will be a useful way of concentrating the minds on clearly defined benefits.

David's postulate is as timely as it is worthwhile. While nobody can expect the muddied waters of the South China Sea to clear up anytime soon, infusing a new dynamism into the current suite of regional economic cooperation programs, if necessary, by completely restructuring them, will define the contrasts between conflict and cooperation. A beginning in changing mindsets might then be possible, and paths to prosperity might well become the roads less traveled.

April 2016
Manila

Arjun Thapan

Preface

This small book reflects my fears that the maritime disputes in the West Pacific Ocean will derail regional development—development that has lifted hundreds of millions of people out of grinding poverty. The on-again, off-again conflicts and disputed claims are over control of passage or to the marine resources, fishing and energy, in the relatively shallow seas. In the north, the disputes are between China, Japan, and South Korea in (using American names) the East China Sea and the Yellow Sea; in the South China Sea they are among Brunei, China, Indonesia, Malaysia, the Philippines, Taiwan, and Viet Nam. The disputes have led to armed, sometimes fatal confrontations and soured international relations.

I take these disputes and the risks to development personally; my wife and I moved first to Asia in the mid-1980s, to Tokyo. There, teaching at Hosei University, I had the opportunity to study firsthand the then rapid growth of the Japanese economy and to travel to and learn about the Chinese economy and the unheralded reforms that were just starting. In the early 1990s, we relocated to the Philippines, where I started work with the Asian Development Bank. For more than two decades, we lived in Manila or Jakarta, watching our children grow and the economies transform.

The process of economic development in East and Southeast Asia is by no means an unalloyed success: many people have been left behind, the environment ravaged, and, with rare exceptions, we have not learned to revise policies and programs that no longer meet the problems of the day. But there has been demonstrable success in raising incomes, lowering mortality rates, easing some of the burdens of poverty, and bringing increased economic opportunity to many. All of this is being put at risk.

Growth in this part of the world is, of course, anchored in the policies of the individual nations, in good governance. Countries that provide education and health services generally do better than others. Governments that carefully invest in infrastructure see a positive return to this effort; conversely, electricity blackouts stifle businesses and growth. But growth was a regional phenomenon—economically rising East Asia owes much to Factory Asia, the web of interwoven global value chains that moves goods between the different countries. This process relies upon open borders, on a peace that seems increasingly fragile.

In spite of the confrontations, I'm convinced that few people actually want the kind of military conflict that might settle the disputes—few people actually call for war. But we may get it. I started writing this book in 2014, the centennial of World War I. That conflict devastated Europe, creating the conditions for the Great Depression and for the catastrophe of World War II. Few people in the years preceding that horrible collective encounter would have guessed the costs they were to pay. Rather, they viewed the expected costs versus the benefits of standing firm and pushing back as acceptable. They were wrong.

One hundred years later we are making the same awful arithmetic mistake; the possible costs of the confrontations are huge, the benefits largely illusionary. A generation after World War I, at the end of World War II, Europe embarked on an alternative path, one of cooperation. As with the earlier American experience of binding disparate groups of people together, the European Union faces continued challenges, but it has proven an alternative to the military conflicts that convulsed Western Europe for decades.

Asia is not likely to follow Europe's path; there are few speaking of economic or political unions. But there are many smaller cooperative initiatives. A plethora of efforts, typically referred to by acronyms, occupy government officials and staff from international organizations in sometimes mind-numbing hours of meetings. Some of these simply provide calls for more meetings. Sometimes, however, they result in changes that improve peoples' lives, expanding economic opportunities and encouraging people to invest in their own economies.

As an economist for the Asian Development Bank I sat through many of those meetings. I helped provide support to a number of regional cooperation initiatives in Southeast Asia. (Also in Central Asia, but that demands another book.) Some of these initiatives helped transform sub-regional economies; some are frustrating "works in progress". The lesson, however, is clear—regional cooperation can sometimes provide an alternative development path, one that can give countries a stake in their neighbors' economic fortunes and reduce the attractiveness of conflict. There are especially opportunities for regional cooperation in the South China Sea, where I focus my attention.

This little book expands on these themes, of the dangers of continued confrontation, of the potential of cooperation. I have many people to thank. From the beginning people read early drafts and helped me think through my arguments, especially Don Uy-barreta, Arjun Thapan, Stephen Groff, Ellen Frost, Douglas Brooks, and Shobhana Murali Stoyanov. My thanks also go to the good people at Palgrave Macmillan for their unflagging encouragement. But finally, I am grateful to my wife who traveled with me and, for so many years, lived with me along the shores of the South China Sea.

Oakland, California, USA David Jay Green
April 2016

Contents

List of Abbreviations and Acronyms

AMTI	Asia Maritime Transparency Initiative
ADB	Asian Development Bank
ADBI	Asian Development Bank Institute
AEC	ASEAN Economic Community
AP	Associated Press
APEC	Asia-Pacific Economic Cooperation
APSC	ASEAN Political-Security Community
ARIC	Asia Regional Integration Center
ASEAN	Association of Southeast Asian Nations
ASEAN DOC	ASEAN Declaration on the Conduct of Parties in the South China Sea
BBC	British Broadcasting Corporation
BIMP-EAGA	Brunei Darussalam-Indonesia-Malaysia-Philippines East ASEAN Growth Area
CFR	Council on Foreign Relations
CMIM	Chiang Mai Initiative Multilateralisation
CNAS	Center for a New American Security
CNOOC	China National Offshore Oil Corporation
CSIS	Center for Strategic and International Studies
EIA	Energy Information Administration (United States)
EAGA	East ASEAN Growth Area
FAO	Food and Agriculture Organization
FDI	Foreign Direct Investment
GDP	Gross Domestic Product
GMS	Greater Mekong Subregion
ICG	International Crisis Group
IMF	International Monetary Fund

IMT-GT	Indonesia-Malaysia-Thailand Growth Triangle
IMS-GT	Indonesia-Malaysia-Singapore Growth Triangle
JTM	Japan Tourism Marketing Co.
Lao PDR	Lao, People's Democratic Republic of
MOFA	Ministry of Foreign Affairs. People's Republic of China
OECD	Organisation for Economic Co-operation and Development
PBG	Pan-Beibu Gulf Economic Cooperation (Gulf of Tonkin)
PRC	People's Republic of China
RCEP	Regional Comprehensive Economic Partnership
Ro-Ro	roll-on/roll-off (ferry systems)
SCS	South China Sea
SIJORI	Singapore-Johor-Riau Growth Triangle
UNCLOS	United Nations Convention on the Law of the Sea
UNCTAD	United Nations Conference on Trade and Development
WTO	World Trade Organization
ZOPFAN	Zone of Peace, Freedom and Neutrality

List of Figures

List of Tables

List of Boxes

1

Introduction

Abstract In the South China Sea disputed claims to ownership of the small islands, to the control of passage, and to the energy and fishery resources are growing sources of tension. Clashes at sea threaten the international trade that has underwritten unprecedented economic growth in East and Southeast Asia. Using simple game theory the book argues that the rapid transformation of the region's economy—the rise of Factory Asia—is not being acknowledged, leading countries to take chances beyond a rational picture of costs and benefits. Regional economic cooperation can be an alternative to the present conflicts. However, while there is the potential for peaceful development of the South China Sea, there are real challenges to structuring successful programs.

Keywords ASEAN • China • Game theory • Philippines • Viet Nam • Regional cooperation

In the western part of the Pacific Ocean, overlapping maritime claims have been the sources of political tension and outright conflict. Brunei, China, Indonesia, Malaysia, the Philippines, Taiwan, and Viet Nam all assert conflicting rights to control of passage in the South China Sea, to the small rocky outcroppings, or to the marine resources.[1] In the East China Sea and the Yellow Sea, similar issues plague relations between China, Japan, and South Korea.[2] At stake are the undersea mineral resources (including possible oil

D.J. Green, *The Third Option for the South China Sea*,
DOI 10.1007/978-3-319-40274-1_1

and natural gas reserves), access to fishing areas (traditionally exploited by vessels from many countries), and the ability to meet security concerns such as controlling the movement of military or commercial vessels.

Figure 1.1 provides a picture of the overlapping maritime claims.[3] While many countries are involved in these conflicts, China predominates.

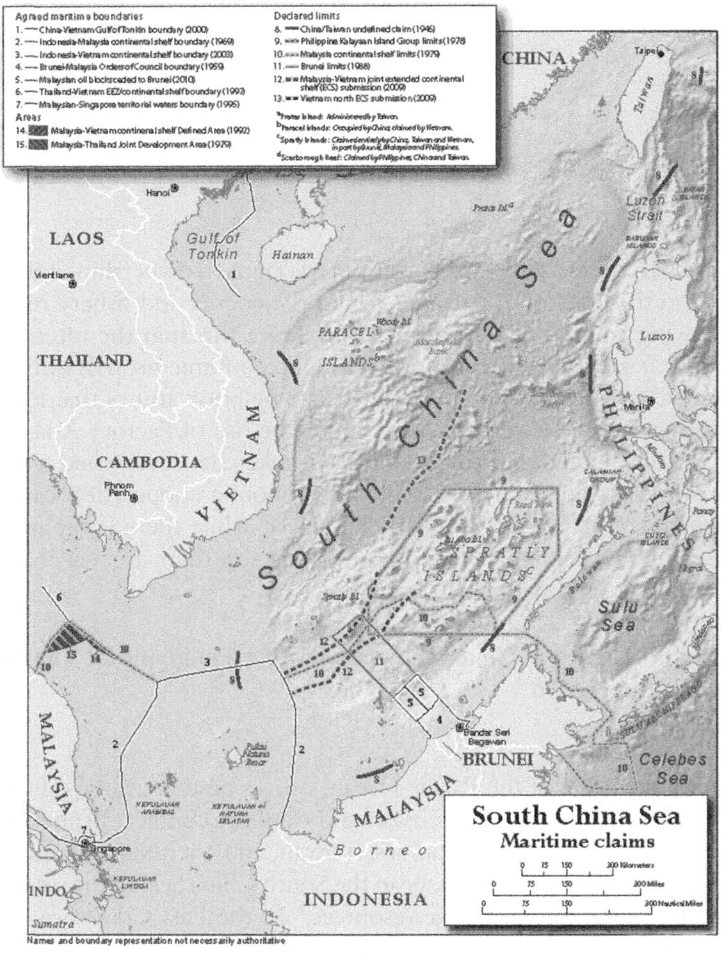

Fig. 1.1 Overlapping maritime claims
Source: US EIA 2013 (http://www.eia.gov/beta/international/analysis_includes/regions_of_interest/South_China_Sea/images/maritime_claims_map.png)

This is a result of a number of factors, including the country's size and geographic centrality, the extent of China's claims, and the explosive growth of the Chinese economy spurring an expansion of both the fishing and naval defense fleets. China's claims follow from those made by the Republic of China in 1947 on the basis of a "nine-dashed line" on a historical map of the region (US EIA 2013). As a result of this history, the Government in Taiwan generally has "mirrored" the claims by China (BBC 2015a), although not completely.[4] The other countries' claims are generally smaller; often claims of exclusive economic zones flowing from their respective coastlines such as provided by the United Nations Convention on the Law of the Sea (UNCLOS).[5]

Table 1.1 provides a summary list of some of these disputes with a short, representative note on the overlapping claims and examples of the violence these have occasioned. A perspective on the political disagreements in the region can be judged by the way the names used for the various bodies of water are contested, see Box 1.1. Although the focus of

Table 1.1 A partial list of and parties to Western Pacific Ocean territorial conflicts

Parties	*Disputed regions* / *Illustrative headline clash or actions*
China-Japan	Diaoyu Islands (Chinese designation)/Senkaku Islands (Japanese designation) in the East China Sea[(1)] 2011, Japanese military aircraft "confront" Chinese plane[(2)] 2012, Chinese naval vessels "blocked" Japanese vessels[(2)]
China-Philippines	The Scarborough Shoal (Huangyan Island in China) and some of the Spratly Islands in the South China Sea[(3)] Naval "standoff" in 2012 at Scarborough Shoal[(4)]
China-South Korea	Maritime resources in Yellow Sea and East China Sea, including a submerged reef, Ieodo (Korean)/Suyan (Chinese)[(5)] 2012 Deadly clashes between Korean maritime security personnel and Chinese fishermen[(5)]
China-Viet Nam	Islands in the Paracel[(6)] and the Spratly[(3)] groups in the South China Sea Violent clashes in 1974 (Battle of the Paracel Islands, between China and South Viet Nam)[(2)] and 1988 (Johnson Reef Skirmish)[(2, 7)] resulting in fatalities; 2014 naval confrontation over Chinese oil drilling rig[(8)]

(*continued*)

Table 1.1 (continued)

Parties	*Disputed regions* / *Illustrative headline clash or actions*
Currently non-violent or less-violent disputes	
Brunei-China	Parts of the maritime territory of Spratly Islands in the South China Sea[9]
China-Indonesia	Region of the South China Sea, including near Indonesian-controlled, natural gas rich, Natuna Islands[10]
	2010 Faceoff between Indonesia naval vessels and "a Chinese 'fishery management vessel'"[11]
China-Malaysia	Maritime area and parts of the Spratly Island group in the South China Sea[3]
	1995, Malaysian naval vessels fire on a Chinese ship[2]
Japan-South Korea	Conflicting claims to South Korea administered Dokdo (Takeshima in Japanese) in the East China Sea[1]
	"Last year (2012) witnessed a particularly angry diplomatic spat, when then South Korea President Lee Myung-bak visited the islands.... Japan responded by recalling its ambassador."[12]
Japan-Taiwan	Tiaoyutai Islands (Taiwanese designation)/Senkaku Islands (Japanese designation) in the East China Sea[2]
	2012 Confrontation between naval vessels[2]
Philippines-Taiwan	The Scarborough Shoal (Huangyan Island in China) and some of the Spratly Islands in the South China Sea[3]
	2013 Fatal encounter between Philippines coast guard and Taiwanese fishing vessel[13]
Philippines-Viet Nam	Conflicting claims to parts of the Spratly Island group in the South China Sea[9]
	1999 Philippines plane fired on by Vietnamese troops[2]; 2011 Naval agreement eased tension[14]
Taiwan-Viet Nam	Islands in the Paracel[6] and the Spratly[3] groups in the South China Sea
	1995 Taiwan military fires upon Vietnamese vessel from Taiwanese held island[2]

Notes: (a) The listing of disputed regions or islands is not meant to be exhaustive or authoritative

(b) The English transliterations of the Chinese and Vietnamese names for the Paracel and Spratly Islands can be found in Buszynski 2011

(1) BBC 2014; (2) CNAS 2014; (3) BBC 2015a; (4) DeCastro 2013; (5) Roehrig 2012; (6) Encyclopedia Britannica 2013; (7) Torode 2013; (8) Spegele and Khanh 2014 and The Economist 2014a; (9) US EIA 2013; (10) Shekhar and Liow 2014; (11) Currie 2010; (12) O'Shea 2013; (13) Thayer 2015; (14) Bordadora 2011

this book is on resolving the South China Sea disputes, the similar issues of the East China Sea need acknowledging—they help us understand the behavioral dynamics, the interrelationships of the countries involved.

Box 1.1 The Name Game The issues over these oceanic swaths on the West Pacific map have been highly politicized, to the extent that the names applied to the different bodies of water have become political issues. (McLaughlin 2011)

The South China Sea is the name most readily recognized in the USA for the body of water centered between China, Taiwan, and the Southeast Asian nations of Brunei, Indonesia, Malaysia, the Philippines, and Viet Nam. This book will use that name. Using English translations, in China, the area is generally referred to as the South Sea, but in Viet Nam, it is the East Sea (Hookway 2013). In the Philippines, it is increasingly referred to as the West Philippine Sea (Green 2013). The Philippine President Benigno Aquino III showed a sense of humor in calling it "this sea known by many names" (Hookway 2013). A US non-governmental organization, the Nguyen Thai Hoc Foundation (2014), advocates using the name Southeast Asia Sea.

Similar issues arise in the west Pacific Ocean areas between China, Japan, and Korea.

Following Baker (2016) we acknowledge that "Maritime borders are ephemeral and subject to diverse concepts of use and passage." But this book will not attempt to further delineate the territorial extent of, the legal aspects surrounding, or the historical development of the overlapping claims to west Pacific Ocean areas. These subjects are well covered in other studies, including Hayton (2014),[6] International Crisis Group (ICG 2012), Cronin (2012), and US EIA (2013); and in a multitude of conferences, some available in published volumes or online.[7]

This book will rather focus on the economic context to and consequences of the South China Sea disputes. In particular, the book will argue that the changing economy of East Asia, especially the flourishing of *Factory Asia* as a web of cross-border investment and trade flows, sharply raises the costs of conflict. Violence in the past, as between China and Viet Nam in 1979, had large economic and human costs (Stout 2014). But at that time, neither country depended principally upon market-based institutions or relied heavily on participation in the global economy, and the economic impacts of these experiences could be contained. It would be dangerous to build expectations of the impact of future conflicts on this historical experience. East Asia owes its sterling economic performance over the last few decades to increasingly interdependent webs of trade and investment flows. Today, real conflict would shake the foundations of Factory Asia.[8]

Emphasizing the economic context is not the only approach that could be taken to examine the South China Sea conflicts; for instance, there is a large literature from the standpoint of national defense or security.[9] It is important not to deny these perspectives: there are many times when national security concerns clearly trump narrow economic worries. However, national security is not a simple concept and understanding a nation's interests demands an awareness of its economic structure and fortunes. Here, the economies of Southeast Asia and China have been changing in a remarkable, unprecedented fashion, altering the costs and benefits of present state behavior. The existing tension and periodic, generally low-level, conflict in Southeast and East Asia have already affected the cross-border investment and trade that virtually defined the process of globalization over the past few decades. This is most evident between China and Japan, but can also be seen between the Southeast Asian disputants and China.

Continued behavior of each government stubbornly standing their ground, or rather line on a sea map, is what is visible and can be considered the current default option for the parties concerned. Experience suggests this is neither likely to lead to a peaceful resolution nor generally to the satisfaction of any particular claimant. Moreover there are risks that the region will slide into open conflict of the type to truly disrupt international trade and investment, imperiling the dramatic gains in economic development over the past few decades and concurrently exacerbating the weak international economy. This second option of regional conflict would be a global disaster.

The current conflicts flow from a variety of historical paths. It is not the goal of this book to establish which side is right. Rather, as the likely costs of conflict are considerable—enough to encourage a search for alternative processes—the book seeks to demonstrate the utility of regional cooperation as a viable alternative to the present behavior. The South China Sea claimants all have experience in large multinational regional cooperation programs. Most particularly the Greater Mekong Subregion (GMS), involving China and Viet Nam, shows the potential for regional cooperation to support inclusive development even with unsettled political issues. Properly constituted, a regional cooperation initiative offers benefits that could well reward the countries for abandoning the present combative postures. While the disputes in the northern and southern portions of the west Pacific Ocean are formally separate, a resolution to the South China Sea disputes through regional cooperation might also change the game in the northwest Pacific, by providing all parties the confidence that there are better options to continued confrontation. The key, however, is to "properly constitute" such an initiative; while the GMS provides positive experiences, other efforts in Southeast Asia show how difficult it is to generate meaningful results.

Chapter 2 of this book provides a sketch of the very variable patterns of violence occasioned by the unresolved maritime claims. These have involved, for instance, confrontations between naval vessels as well as the building of facilities on islands and semi-submerged reefs in the South China Sea. Chapter 3 suggests that a game theory framework can explain this uneven and changing pattern of low-level conflict, in particular that each contestant sees only an occasional need to aggressively assert national claims. The conclusion of this section is, however, that there is a real probability of sliding into true conflict. Chapter 4 argues that the costs of such a conflict are likely larger than currently perceived and, due to growing regional economic interrelationships, increasing, raising the stakes for finding alternative behavior. The book emphasizes the importance of interlinked flows of international trade, but also the crucial value of cross-border investment, especially given the evolving Chinese economy and its need for targets for outward foreign investment. If the costs and risks of continued conflict are large, Chap. 5 concludes that the narrow economic benefits of "winning" any conflict—the control of mineral and fishery resources—are small. Chapter 6 provides a perspective on the earlier discussion by introducing a broader set of concerns, including those related to security, that motivate national behavior with respect to the South China Sea. This section also reviews the interests and impact of non-regional players, particularly the USA. Chapter 7 introduces into the game theory model the possibility of regional cooperation as a third alternative to the present behavior of intermittent passive and aggressive pursuit of rival maritime claims, concluding that regional cooperation must be more than simply sharing existing resources; the exercise must provide some real additionality to the current contest. Chapter 8 reviews the Southeast Asian experience in regional cooperation initiatives, drawing lessons for this alternative. The concluding Chaps. 9 and 10 argue that there indeed could be a third option for the South China Sea, regional collaboration leading to quickened economic growth and development, but this is not simple to find or initiate. The conclusion thus poses a challenge; while the effort needed to successfully mount a viable regional cooperation program is formidable, the risks to the regional and indeed global economy are too large to allow the present context and behavior to go unchanged.

Notes

1. There are a large number of studies and other sources of information focusing on the western Pacific disputes, including the website of the Center for a New American Security (CNAS 2014), Hayton (2014), ICG (2012), Rosenberg's website (Rosenberg SCS online), and Xu (2014). Much of this work, such as that sponsored by the Center for Strategic and International Studies (Cronin 2013), concentrates on the strategic concerns associated with the South China Sea.
2. The conflicts addressed in this book are by no means the only ones in the western Pacific. For instance, in some areas, Russia and Japan face off and the Korean peninsula conflict remains unresolved. For purposes of simplicity, this book will simply refer to the East China Sea when discussing that body of water as well as to the Yellow Sea. In the text, countries will generally be listed alphabetically.
3. There are many instructive maps that show the overlapping claims, especially in the area of the Spratly Islands. See the New York Times website (New York Times 2012) that depicts the myriad of islands. Another is Prof. David Rosenberg's website. (Rosenberg, SCS online).
4. Shu-Ling (2013) suggests that a pact signed with Japan by Taiwan over fishing rights subtly distinguished Taiwan's policies from China. The Economist (2015) also argues that Taiwan's claims are "narrower" than pursued by the mainland government.
5. The U.S. Energy Information Administration (EIA 2013) provides admirably concise tabular data on the overlapping claims.
6. Hayton (2014) provides a very intriguing account of the history of the South China Sea, emphasizing that in Southeast Asia, generally the concept of borders is a recent transplant, helping explain the present lack of clarity as to who owns what. However, Hayton also notes that the current positioning over ownership of the rocky islets dates at least to the beginning of the last century.
7. For example, Thuy and Trang (2015), and see the conferences hosted by the Center for a New American Security (Cronin 2012) and the Center for Strategic and International Studies (CSIS 2015).
8. This book is certainly not alone in arguing that the failure to resolve these maritime conflicts could be hugely damaging to the region—most studies at least seem to tacitly presume this. In a slightly different vein, Rogers (2012, p. 85) explicitly suggests that countries involved in disputes may not fully take into account changing trends, especially related to energy resources,

and thus may "misinterpret the actions of their neighbors and the risk of instability may increase."

9. See, for instance, CSIS (2016) or Shear (2015).

Bibliography

Baker, Rodger. 2016. Fish: The overlooked destabilizer in the South China Sea, 12 Feb. 2106. Stratfor. https://www.stratfor.com/analysis/fish-overlooked-destabilizer-south-china-sea. Accessed 21 Mar 2016.

Bordadora, Norman. 2011. Philippines, Vietnam forge naval agreement on Spratlys. Philippine Daily Inquirer, 27 Oct 2011. http://globalnation.inquirer.net/16517/philippines-vietnam-forge-naval-agreement-on-spratlys. Accessed 21 June 2015.

British Broadcasting Corporation (BBC). 2014. How uninhabited islands soured China-Japan ties. BBC News, 10 Nov 2014. http://www.bbc.com/news/world-asia-pacific-11341139. Accessed 16 Dec 2014.

———. 2015a. Q&A: South China Sea dispute. BBC News, 27 Oct 2015. http://www.bbc.com/news/world-asia-pacific-13748349. Accessed 22 July 2015.

Buszynski, Leszek. 2011. The South China Sea: Avenues towards a resolution of the issue. Southchinaseastudies.org, 24 Mar 2011. http://southchinaseastudies.org/en/conferences-and-seminars-/515-the-south-china-sea-avenue-towards-a-resolution-of-the-issue. Accessed 21 May 2015.

Center for a New American Security (CNAS). 2014. Timeline: 1955-Present. Cnas.org. http://www.cnas.org/flashpoints/timeline. Accessed 3 Jan 2015.

Center for Strategic and International Studies (CSIS). 2015. 21 Jul 2015, Washington, DC. http://csis.org/event/fifth-annual-csis-south-china-sea-conference. Accessed 16 Feb 2016.

———. 2016. Asia-Pacific rebalance 2025 capabilities, presence, and partnerships: An independent review of U.S. defense strategy in the Asia-Pacific. http://csis.org/files/publication/160119_Green_AsiaPacificRebalance2025_Web_0.pdf. Accessed 11 Mar 2016.

Cronin, Patrick M. (ed). 2012. Cooperation from strength: The United States, China and the South China Sea. Washington, DC: Center for a New American Security. http://www.cnas.org/sites/default/files/publications-pdf/CNAS. Accessed 16 May 2015.

——— 2013. The strategic significance of the South China Sea. For "Managing tensions in the South China Sea" conference held by the Center for Strategic & International Studies (CSIS) on 5–6 June 2013. http://csis.org/files/attachments/130606_Cronin. Accessed 23 Nov 2014.

Currie, Kelley. 2010. Why is China picking fights with Indonesia? The Weekly Standard. http://www.weeklystandard.com/blogs/why-china-picking-fights-indonesia. Accessed 4 Jan 2015.

De Castro, Renato Cruz. 2013. China's realpolitik approach in the South China Sea dispute: The case of the 2012 Scarborough Shoal stand-off. For Managing Tensions in the South China Sea conference held by the Center for Strategic & International Studies (CSIS) on 5–6 June 2013. http://csis.org/files/attachments/130606_DeCastro_ConferencePaper.pdf. Accessed 23 Nov 2014.

The Economist. 2014a. Not the usual drill. 10 May 2014. http://www.economist.com/news/asia/21601879-tensions-mount-dangerously-contested-waters-not-usual-drill. Accessed 17 Nov 2014.

———. 2015. Small reefs, big problems. 25 Jul 2015. http://www.economist.com/news/asia/21659771-asian-coastguards-are-front-line-struggle-check-china-small-reefs-big-problems. Accessed 1 Aug 2015.

Encyclopedia Britannica. 2013. Paracel Islands | islands, South China Sea. Encyclopedia Britannica, 21 Jul 2013. http://www.britannica.com/EBchecked/topic/442423/Paracel-Islands. Accessed 3 Jan 2015.

Green, David Jay. 2013. Fighting over the West Philippine Sea is so 17th century. *Business Mirror*, Manila, 30 June 2013.

Hayton, Bill. 2014. *The South China Sea: The struggle for power in Asia*. New Haven: Yale University Press.

Hookway, James. 2013. What's in a name? In the South China Sea, it seems, quite a lot. *The Wall Street Journal*, 9 Oct 2013. http://blogs.wsj.com/indonesiarealtime/2013/10/09/whats-in-a-name-in-the-south-china-sea-it-seems-quite-a-lot/?mg=blogs-wsj&url=http%3A%2F%2Fblogs.wsj.com%2Fsearealtime%2F2013%2F10%2F09%2Fwhats-in-a-name-in-the-south-china-sea-it-seems-quite-a-lot%2F. Accessed 16 Dec 2014.

International Crisis Group (ICG). 2012. Stirring up the South China Sea (II): Regional responses, report N 229, 24 July 2012. http://www.crisisgroup.org/~/media/Files/asia/north-east-asia/229-stirring-up-the-south-china-sea-ii-regional-responses. Accessed 22 May 2015.

McLaughlin, Kathleen E. 2011. South China Sea: China vs. Vietnam. *GlobalPost*, 14 Jun 2011. http://www.globalpost.com/dispatch/news/regions/asia-pacific/china/110614/south-china-sea-china-vs-vietnam. Accessed 20 Jan 2015.

Nguyen Thai Hoc Foundation. 2014. Change the name "South China Sea" to "Southeast Asia Sea", Change.org. https://www.change.org/p/change-the-name-south-china-sea-to-southeast-asia-sea. Accessed 16 Dec 2014.

New York Times. 2012. Territorial claims in South China Sea. Nytimes.com, 31 May 2012. http://www.nytimes.com/interactive/2012/05/31/world/asia/Territorial-Claims-in-South-China-Sea.html?_r=0. Accessed 3 Jan 2015.

O'Shea, Paul. 2013. Territorial disputes in Northeast Asia: A primer. Italian Institute for International Political Studies Analysis (ISPI) No. 182, June 2013. http://www.ispionline.it/sites/default/files/pubblicazioni/analysis_182_2013.pdf. Accessed 8 Dec 2014.

Roehrig, Terence. 2012. South Korea–China maritime disputes: Toward a solution. 27 Nov 2012. East Asia Forum. http://www.eastasiaforum.org/2012/

11/27/south-korea-china-maritime-disputes-toward-a-solution/. Accessed 1 Dec 2014.

Rogers, Will. 2012. The role of natural resources in the South China Sea. In *Cooperation from strength the United States, China and the South China Sea*, ed. P. Cronin, 83–97. Washington, DC: Center for a New American Security. http://www.cnas.org/sites/default/files/publications-pdf/CNAS. Accessed 16 May 2015.

Rosenberg, David. (SCS online). The South China Sea. http://www.southchinasea.org/. Accessed 17 May 2015.

Shear, David. 2015. Statement of David Shear Assistant Secretary of Defense for Asian & Pacific Security Affairs before the Senate Committee on Foreign Relations, May 13, 2015. U.S. Senate Committee on Foreign Relations. http://www.foreign.senate.gov/imo/media/doc/051315_Shear_Testimony.pdf. Accessed 6 Mar 2016.

Shekhar, Vibhanshu, and Joseph Chinyong Liow. 2014. Indonesia as a maritime power: Jokowi's vision, strategies, and obstacles ahead. The Brookings Institution. Nov 2014. http://www.brookings.edu/research/articles/2014/11/indonesia-maritime-liow-shekhar. Accessed 16 Dec 2014.

Shu-Ling, Ko. 2013. Details of new Japan-Taiwan fisheries pact are explained. The Japan Times, 23 Apr 2013. http://www.japantimes.co.jp/news/2013/04/23/national/details-of-new-japan-taiwan-fisheries-pact-are-explained/#.VVwhCGTBzGc. Accessed 20 May 2015.

Spegele, Brian, and Vu Trong Khanh. 2014. China, Vietnam try to repair ties after South China Sea dispute. *The Wall Street Journal*. 28 Aug 2014. http://www.wsj.com/articles/china-vietnam-try-to-repair-ties-after-oil. Accessed 11 Jan 2015.

Stout, David. 2014. The last time China got into a fight with Vietnam, it was a disaster, 15 May 2014. TIME. http://time.com/100417/china-vietnam-sino-vietnamese-war-south-china-sea/. Accessed 17 Apr 2015.

Thayer, Carlyle A. 2015. South China Sea tensions: China, the claimant states, ASEAN and the major powers. In *Power, law, and maritime order in the South China Sea*, ed. T.T. Thuy and L.T. Trang, 3–35. Lanham: Lexington Books.

Thuy, Tran Truong, and Le Thuy Trang (eds.). 2015. *Power, law, and maritime order in the South China Sea*. Lanham: Lexington Books.

Torode, Greg. 2013. Spratly Islands dispute defines China-Vietnam relations 25 years after naval clash. South China Morning Post, 23 Dec 2014. http://www.scmp.com/news/asia/article/1192472/spratly-islands-dispute-defines-china-vietnam-relations-25-years-after. Accessed 2 Jan 2015.

United States Energy Information Administration (US EIA). 2013. South China Sea. http://www.eia.gov/beta/international/regions-topics.cfm?RegionTopicID=SCS. Accessed 23 Jul 2015.

Xu, Beina. 2014. South China Sea tensions. Council on Foreign Relations—CFR Backgrounders, 14 May 2014. http://www.cfr.org/china/south-china-sea-tensions/p29790. Accessed 22 May 2015.

2

Low-Level Simmering Disputes

Abstract This chapter of this book provides a sketch of the very variable patterns of violence occasioned by the unresolved maritime claims. These have involved, for instance, confrontations between naval vessels as well as the building of facilities on islands and semi-submerged reefs, especially in the South China Sea. The discussion raises the question, to be answered in the following chapter, "why do we see very uneven variations between aggressive and passive behavior in support of conflicting claims by all parties in the dispute?" The most recent time period, involving reclamation of rocky features and small islands, is examined and seen as a dangerous militarization of the South China Sea.

Keywords Conflict • South China Sea • China • Viet Nam • Philippines • Spratly Islands

2.1 A Long Period of Intermittent Confrontations

Figure 2.1 provides a sense of the level of conflict in the South China Sea and the analogous picture for the East China Sea. The bulk of the data were taken from a timeline, a list of reported clashes or threatening or aggressive actions by the rival maritime claimants in these two areas of

D.J. Green, *The Third Option for the South China Sea*,
DOI 10.1007/978-3-319-40274-1_2

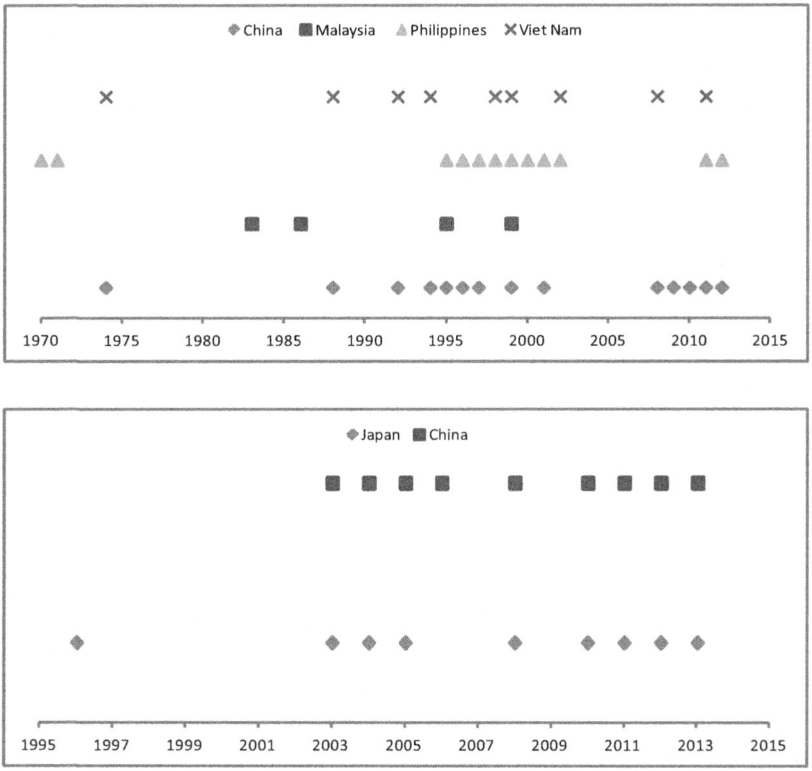

Fig. 2.1 Reports of aggressive behavior in the Western Pacific Ocean. *Notes*: The data are on an annual basis with most of the reports from the Center for a New American Security (CNAS 2014). Incidents were noted if they appeared to signal an aggressive posture to assert control of maritime area. Additional observations were added from various published reports, particularly with respect to "island building"—the enlargement of small outcroppings. The observations are annual through 2013 and could mask multiple actions. The vertical arrangement, which country is higher than any other, is arbitrary

the western Pacific Ocean, collected by the Center for a New American Security. (CNAS 2014) The dataset ended in 2013 and was supplemented by other reports, for instance, of one country occupying and enlarging the small, generally uninhabited islets as a way of occupying marine space. As such, however, the picture does not adequately represent the tensions over the last few years. This is addressed separately below.

The translation of the raw news into the spots on Fig. 2.1 was done on a simple dichotomous basis—any report of apparently purposeful violent or threatening actions was listed as an instance of aggressive behavior; multiple reports in a given year were taken as confirming evidence of a basic stance but not otherwise noted.[1] For instance, a report that a military vessel of country X was threatening a fishing boat of country Y would earn country X a mark for aggressive behavior; a standoff between two naval vessels would earn both countries a mark.

Other regional economies such as Indonesia, South Korea, and Taiwan and non-regional countries including India and the USA have also been reported as being involved in violent or threatening incidences, but to a much smaller degree and have not been included in that picture. This can change. Recently in Indonesia, the Government has taken to more aggressively defending its maritime areas from foreign fishing boats. (Associated Press [AP] 2015) This would involve Indonesian security vessels in what other countries would consider to be aggressive behavior if it occurs in disputed areas. In another example, in October 2015, the USA sent a warship close to a Chinese claimed islet in the South China Sea. (Lubold and Page 2015) Involvement of non-regional players clearly has the potential for changing the nature of the disputes and their impact on international relations and the global economy.

Complicating any analysis, what is considered *aggressive* in one period of time might not be seen that way in another. For instance, currently considerable attention is being paid to the occupation of islands or rocky islets in the South China Sea, but many of these had been occupied at different periods of time without necessarily resulting in the same level of international tension. The CNAS data set clearly makes a good attempt to record aggressive behavior by the different countries. It is likely that every incident noted would be so regarded by one of the countries as aggressive behavior, but it is also likely that not every aggressive act has been reported or noted.[2] For example, the International Crisis Group (2012, p. 10, fn. 94) reports "confrontations between Malaysian fishing boats and Indonesian vessels have involved high-calibre weapons." This was not noted in the CNAS data set. It should be clear that this analysis is both subjective and partial at best. It is meant to frame the argument, not strictly define it.

The underlying reports were always bilateral—in all cases, only two countries were involved. The bottom portion of Fig. 2.1, being limited to China and Japan, reflects their confrontations, but otherwise no attempt has been made to delineate the pairing of incidents. Overall, there are disputes noted between the different claimants, for instance between the Philippines and Viet Nam, but China has been involved in more reported events than other countries. Partly this simply reflects the large overlap between Chinese maritime claims and that of rival countries.

Table 2.1 summarizes one aspect of these figures: the frequency of observed aggressive behavior over the time periods considered. The beginning dates for these two time periods are arbitrary, trying simply to portray behavior in decades in which there are incidents of aggressive behavior noted in the CNAS database. The Table suggests considerable, low-level conflict in the South China Sea. Within the scope of the database, during 1970–2013, China has engaged in aggressive assertion of its claims in 32 % of the years, and the Philippines 27 %.[3]

There are many questions, but one that begs to be answered is, "Why was the pattern of incidents or aggressive behavior in Fig. 2.1 so uneven, so patchy?" In the South China Sea, there does seem to be a pick-up in the 1990s, especially visible with the actions of China and the Philippines. Fravel (2012, p. 33) argues that the region exhibited more "instability" between 1988 and 1995 than seen later. Soon Ho Lee (2013) suggests that some aspect of this may reflect the UN Convention on the Law of the Sea (UNCLOS) which encouraged governments to make known their claims in the lead up to 1994 when it became operational.[4] While there is likely to be a variety of proximate causes for particular incidents, we will argue below that the pattern of on-again, off-again violent or threatening actions suggests an underlying uncertainty about the approach to be taken to attain these goals.[5] Some insight into this aspect of these conflicts can be had using game theory.

Table 2.1 Frequency of observed aggressive behavior

South China Sea	**Time period 1970–2013**
China	32 %
Malaysia	9 %
Philippines	27 %
Viet Nam	20 %
East China Sea	**Time period 1990–2013**
Japan	38 %
China	38 %

Notes: The frequency of annual reports of aggressive behavior by each country in the given time period. See Fig. 2.1

2.2 Today's Conflict: Building and Rebuilding Islands in the Sea

However, before dealing with this central issue as to how to explain the on-again, off-again aggressive behavior seen in the South China Sea, we need to look at the events of the last few years. During this most recent period, we have seen a definite uptick in tension, especially reflecting reports of Chinese construction or expansion of permanent facilities on different parts of the contested maritime region. Sometimes this activity involves building on what had hitherto been uninhabited, sometimes submerged reefs.

As mentioned earlier, occupying, building structures on the islands is not new. Shear (2015, p. 2) comments: "In the Spratly islands, Vietnam has 48 outposts; the Philippines, 8; China, 8; Malaysia, 5, and Taiwan, 1."[6] The examples below provide some of the flavor of a complex picture of different countries, each establishing a presence over several decades in the Spratly Islands group in the South China Sea:

- Taiping Island, occupied "since the 1940s," has an airport (Jennings 2013) and saw troops stationed by Taiwan in 1956. (CNAS 2014)
- The Philippines also maintains an airfield on Thitu Island (referred to as Pagasa Island in the country). The island, occupied for decades and administered as part of the Filipino local government systems, appears to have received much less by way of capital investment—the airfield is, for example, unpaved. (Mogato 2015)
- Viet Nam occupied Sand Cay in 1975, and the island has seen, since 2011, land reclamation, expanding the original size by "more than 50 %, adding extensive defensive structures and facilities" (AMTI 2015b).
- Malaysia has expanded Swallow Reef (occupied in 1983) to include an airstrip and a dive resort (Quintos 2015).
- Fiery Cross Reef, occupied by the Chinese since 1987, is the site of "reclamation [that] has increased the area ... 11 times over" with the principal land enlargement beginning in 2014 and extensive construction of facilities including three cement plants (AMTI 2015a). Although somewhat difficult to grasp, given the current context, the original occupation and construction by the Chinese Government was said to be in order to fulfill a commitment to UNESCO for tracking weather (AMTI 2015a).

All countries occupying islands or other features in the South China Sea have engaged in some sort of building or construction. But the relatively recent efforts of the Chinese Government have dwarfed those of their competitors (Shear 2015). The Asia Maritime Transparency Initiative of the Center for Strategic and International Studies estimates that, while Viet Nam has expanded two islands (West Reef and Sand Cay) by 86,000 square meters, the Chinese work on Fiery Cross Reef alone transformed a semi-submerged feature into an artificial island that covers 900,000 square meters. (AMTI 2015c) In addition to the work on Fiery Cross Reef, major Chinese construction efforts have been seen on more than half a dozen other features in the Spratlys.[7] (AMTI 2015b). The contrast between existing "dilapidated" facilities on the Filipino-controlled Thitu (Mogato 2015) and the reported extensive building on the Chinese-occupied islands—including a 3110-meter airfield on Fiery Cross Reef (AMTI 2015b)—is telling.

There are a number of reasons for these differences between the Chinese building and investment effort in the Spratlys and those of the other countries. One is simply the timing: the Filipino, Malaysian, Taiwanese, and Vietnamese governments took control of their islands decades ago when many were unoccupied. The Chinese Government, in the Spratlys, in many cases, elected to make do with semi-submerged reefs or small islets, requiring considerable investment just to safely house people. Tran Truong Thuy (2015) characterizes this difference as "first come, first build."[8] However, it is clear that the Chinese effort is designed to do more than simply provide safe harbors. Thuy (2015) further suggests that one reason for the extensive and expensive efforts is to send the message that this part of the sea is "legally" under Chinese Government jurisdiction, as much as any coastal area would be.

The nature of the Chinese Spratly building program is also designed to send more than a message in support of its claims to the maritime region—many of the facilities provide for security forces. This has been widely described. Erickson and Strange (2014), for instance, note that on Johnson South Reef alone, over a two-year period before 2014, "China appears to have set up additional radars, satellite communication equipment, anti-aircraft and naval guns, a helipad, a dock, and even a wind turbine."[9] In contrast, especially the Philippine-controlled areas appear to have little in the way of support for their military.[10] However, the Philippine Government may well respond to the Chinese building program with one of its own, as may the other governments. Indeed, all disputants appear to be increasing their spending on the military capability to assert their maritime claims. (Chang 2014)

In the context of this book, as discussed below, the island-building effort both facilitates and encourages undertaking aggressive behavior. To see this, we first develop an analytic model, based on game theory, that helps explain the behavior of the rival claimants, especially that their behavior can be expected to vary between passive and aggressive confrontations.

Notes

1. Aggressive behavior is distinct from conflict; compare Storey's (2012, p. 58) "coercive pressure". While this section of the book simply divides behavior into "aggressive" and "passive," reality is different. For example, Holmes (2012, p. 110) makes the case that China's use of "enforcement vessels rather than naval vessels" to impose fishing regulations on boats of other nations, while clearly aggressive, sends a nuanced message that the South China Sea should be considered equivalent to coastal, territorial waters.
2. Fravel (2012, p. 43) provides a different timeline of actions by countries disputing China's claims to the South China Sea, including, for instance, legislative actions by the different governments.
3. While this framework is useful in illustrating the broader argument, it has to be noted that using "annual" frequencies is completely arbitrary.
4. Buszynski (2011) argues that the influence of UNCLOS emerged much earlier as the negotiation started in 1973.
5. Others have noted that the behavior of the participants in these conflicts appears to have inconsistent features or aspects that vary over time. Bader et al. (2014, p. 3) characterize the level of conflict in the South China Sea as "occasional assertions of sovereignty by one actor or another [that] are not a fundamental challenge to recognized borders or the integrity of existing states."
6. This kind of counting and the underlying situation is very fluid and different authors provide different numbers. See, for instance, Quintos (2015, p. 2). As another example, as the manuscript for this book was being finalized, there were notices in the Philippine press in early March 2016 that China had sent naval ships to the Jackson or Quirino Atoll, perhaps to begin an occupation, and that these ships had been blocking Filipino fishing boats from working in the area. Later it was reported that these vessels had left the area. (Calleja 2016)
7. That these activities also have a very large negative impact on the shallow sea environment has not gone unnoticed. See Stoa (2015) and CSIS (2016, p. 74).
8. The CSIS (2016, p. 15) also suggests that the very vigorous island-building campaign is an effort to catch-up to the earlier efforts by rival South China Sea claimants.

9. The authors speculate that the larger effort on Fiery Cross would create "a military base twice the size of Diego Garcia, a key U.S. military base in the Indian Ocean" Erickson and Strange (2014).
10. Compare the extensive references to military support facilities on the Chinese and Vietnamese controlled islands with Mogato's (2015) description of the Philippine-controlled Tithu: "The only sign that the island hosts a military base are two 40 mm anti-aircraft guns on opposite sides of the runway."

Bibliography

Asia Maritime Transparency Initiative (AMTI). 2015a. A fiery cross to bear. http://amti.csis.org/fiery-cross/. Accessed 4 Mar 2016.

———. 2015b. Island tracker. http://amti.csis.org/island-tracker/. Accessed 4 Mar 2016.

———. 2015c. Sandcastles of their own: Vietnamese expansion in the Spratly Islands. http://amti.csis.org/vietnam-island-building/. Accessed 4 Mar 2016.

Associated Press (AP). 2015. Indonesia sinks 41 foreign boats to warn against poaching. Yahoo News, May 20, 2015. http://news.yahoo.com/indonesia-sinks-41-foreign-boats-warn-against-poaching-144052862.html. Accessed 25 Mar 2016.

Bader, Jeffrey, Kenneth Lieberthal, and Michael McDevitt. 2014. Keeping the South China Sea in perspective. Brookings, August 2014. http://www.brookings.edu/research/papers/2014/08/south-china-sea-perspective-bader-lieberthal-mcdevitt. Accessed 20 May 2015.

Buszynski, Leszek. 2011. The South China Sea: Avenues towards a resolution of the issue. Southchinaseastudies.org, 24 Mar 2011. http://southchinaseastudies.org/en/conferences-and-seminars-/515-the-south-china-sea-avenue-towards-a-resolution-of-the-issue. Accessed 21 May 2015.

Calleja, Niña P. 2016. Chinese vessels have left Quirino Atoll—DFA. Philippine Daily Inquirer, 2 Mar 2016. http://globalnation.inquirer.net/137219/chinese-vessels-have-left-quirino-atoll-dfa-dnd. Accessed 6 Mar 2016.

Center for a New American Security (CNAS). 2014. Timeline: 1955-Present. Cnas.org. http://www.cnas.org/flashpoints/timeline. Accessed 3 Jan 2015.

Center for Strategic and International Studies (CSIS). 2016. Asia-Pacific rebalance 2025 capabilities, presence, and partnerships: An independent review of U.S. defense strategy in the Asia-Pacific. http://csis.org/files/publication/160119_Green_AsiaPacificRebalance2025_Web_0.pdf. Accessed 11 Mar 2016.

Chang, Felix K. 2014. Comparative Southeast Asian military modernization—1. The Asan Forum: Topics of the month, 01 Oct 2014. http://www.theasanforum.org/comparative-southeast-asian-military-modernization-1/. Accessed 18 May 2015.

Erickson, Andrew S., and Austin Strange. 2014. Pandora's Sandbox: China's Island-Building Strategy in the South China Sea. Foreign Affairs. 13 Jul. 2104. https://www.foreignaffairs.com/articles/china/2014-07-13/pandoras-sandbox. Accessed 6 Mar. 2016.

Fravel, M. Taylor. 2012. Maritime Security in the South China Sea and the Competition over Maritime Rights. In: P. Cronin, ed., Cooperation from Strength The United States, China and the South China Sea. Washington, DC: Center for a New American Security. pp.31–50. http://www.cnas.org/sites/default/files/publications-pdf/CNAS_CooperationFromStrength_Cronin_1.pdf. Accessed 16 May 2015.

Holmes, James R. 2012. Rough Waters for Coalition Building. In: P. Cronin, ed., Cooperation from Strength The United States, China and the South China Sea. Washington, DC: Center for a New American Security, pp. 99–115. http://www.cnas.org/sites/default/files/publications-pdf/CNAS_CooperationFromStrength_Cronin_1.pdf. Accessed 16 May 2015.

International Crisis Group (ICG, 2012). Stirring up the South China Sea (II): Regional Responses, Report N°229, 24 July 2012. http://www.crisisgroup.org/~/media/Files/asia/north-east-asia/229-stirring-up-the-south-china-sea-ii-regional-responses. Accessed 22 May 2015.

Jennings, Ralph. 2013. South China Sea: Taiwan Snaps back as big fish circle. Forbes, 3 Sep 2013. http://www.forbes.com/sites/ralphjennings/2013/09/03/tai-wan-snaps-back-as-bigger-sh-circle/#1b96f5452753. Accessed 5 Mar 2016.

Lee, Soon Ho. 2013. Japan's dilemma: Maritime disputes in East Asia. East Asia Forum, 2 Mar 2013. http://www.eastasiaforum.org/2013/03/02/japans-dilemma-maritime-disputes-in-east-asia/. Accessed 29 Jan 2015.

Lubold, Gordon, and Jeremy Page. 2015. U.S. Navy Ship Sails Near Islands Claimed By China. The Wall Street Journal. 26 Oct. 2015. http://www.wsj.com/articles/u-s-navy-ship-sails-near-islands-claimed-by-china-1445908192?alg=y. Accessed 30 Oct. 2015.

Mogato, Manuel. 2015. From rundown outpost, Philippines watches China Island take shape in disputed sea. Reuters. 11 May 2015. http://www.reuters.com/article/us-southchinasea-philippines-outpost-idUSKBN0NW18F20150511. Accessed 5 Mar 2016.

Quintos, Mary Fides A. 2015. Artificial Islands in the South China Sea and their Impact on Regional (In)security. FSI Insights, Center for International Relations and Strategic Studies (CIRSS) of the Foreign Service Institute (FSI), Mar. 2015 (II/#2). http://www.fsi.gov.ph/wp-content/uploads/2015/03/2015-0305-Vol-2-No-2-FSI-Insights-Artificial-Islands-in-the-South-China-Sea-Quintos.pdf. Accessed 5 March 2016.

Shear, David. "Statement of David Shear Assistant Secretary of Defense for Asian & Pacific Security Affairs before the Senate Committee on Foreign Relations, May 13, 2015". U.S. Senate Committee On Foreign Relations. http://www.foreign.senate.gov/imo/media/doc/051315_Shear_Testimony.pdf. Accessed 6 Mar. 2016.

Stoa, Ryan. 2015. Environmental Peacebuilding in the South China Sea. 1 & 9 May 2015 [Blog] http://ryanstoa.com/blog?category=South+China+Sea. Accessed 25 Mar. 2016.

Storey, Ian. 2012. China's Bilateral and Multilateral Diplomacy in the South China Sea. In: P. Cronin, ed., Cooperation from Strength The United States, China and the South China Sea. Washington, DC: Center for a New American Security, pp. 51–66. http://www.cnas.org/sites/default/files/publications-pdf/CNAS_CooperationFromStrength_Cronin_1.pdf. Accessed 16 May 2015.

Thuy, Tran Truong. 2015. Construction in the South China Sea: A comparative view. Asia Maritime Transparency Initiative. http://amti.csis.org/construction-in-the-south-china-sea-a-comparative-view/. Accessed 5 Mar 2016.

3

A Model from Game Theory

Abstract This chapter suggests that the game of *chicken*, a simple model from game theory, can explain the uneven and changing pattern of low-level conflict described in Chap. 2. The simple model is used to (i) explain why countries, using "mixed strategies," would alternate between aggressive and non-aggressive behavior in support of their maritime claims; (ii) draw attention to the need to more fully understand the costs and benefits of alternative scenarios; and (iii) motivate discussion about whether the conflict in the South China Sea can only be seen as non-cooperative or whether there are cooperative solutions to the problem. One conclusion is that there is the real risk of sliding into a regional conflict.

Keywords Game theory • South China Sea • Conflict • Regional and international cooperation

The formal structure of game theory helps us see why it is sensible that countries alternate between aggressive and passive behaviors.[1] While sensible in the small, artificially contained picture of game theory, this further suggests that there is no stable macro-equilibrium—that there are risks the countries involved will slide into intensified conflict with the human costs that war can bring as well as disrupting the international economy.

Simplifying the real situation, we restrict ourselves to an uncomplicated, bilateral game that is played on a repetitive basis; and assume that the two

D.J. Green, *The Third Option for the South China Sea*,
DOI 10.1007/978-3-319-40274-1_3

countries see only two alternatives, either aggressively asserting territorial claims or behaving passively, which might allow the other to have control. We assume further that

- cooperation is not feasible, each country behaves in an independent fashion, taking action at the same time;
- there are few or no rewards to yielding to the opposing party;
- conversely, the gains to realizing some exclusive territorial use are substantial;
- the costs and benefits stay constant over time; and
- the costs of mutual defiance, of contemporaneous aggressive behavior, are significant.

Table 3.1 illustrates this game. In this structure, countries A and B have conflicting claims to the maritime region with its resources and two alternative behaviors, either to (i) passively or (ii) aggressively assert their claims. This yields a 2 × 2 game. In one case (upper left corner), if both countries behave passively or non-aggressively in support of their individual claims there would be some benefits from non-exclusive resource use for each country. For instance, bilateral passive behavior would allow fishing boats from either country to ply their trade and perhaps for the two countries, in parallel, to exploit the mineral resources. In the two cells that reflect one country behaving in an aggressive fashion and one country not, it is assumed that there would be considerable and asymmetric benefits

		Country A	
		Passively assert claims	Aggressively assert claims
Country B	Passively assert claims	parallel resource exploitation parallel resource exploitation	gain exclusive territorial rights loss of any resources
	Aggressively assert claims	loss of any resources gain exclusive territorial rights	loss of resources + open conflict loss of resources, + open conflict

Table 3.1 A game theory picture of South China Sea disputes

and costs. The country behaving aggressively would, for the given period, win exclusive rights to some of the resources, such as fishing rights or control of some aspect of maritime transit. The passive party would see no rewards. Finally, there is the lower right-hand cell that reflects the costs of both countries refusing to recognize the claims of the other. In this case, we imagine at least limited actual conflict with immediate opportunity costs to both parties of the inability to exploit the maritime region as well as the costs of conflict and those due to the disruption of international trade and cross-border investment.

Using the sharply narrow structure of a game is bound to have costs in a failure to reflect the actual complexity of the real world of international relations. The use of a 2 × 2 matrix to represent the very complex, multilateral set of interactions involving a wide range of decision-makers, each attempting to work toward many different, even conflicting, goals, operating with uncertain information in a shifting environment, is a heroic set of assumptions. The structure was chosen to be reasonably compatible with some important aspects of the observed current behavior, in particular, the alternatively passive-aggressive behavior. More broadly, the use of the model (i) represents some salient features of observed behavior, that conflicts have been bilateral and that they are justified with reference to resources or control of transit; (ii) highlights the risks of the current situation, including the risk of real, development-destroying conflict; and (iii) provides a platform to examine alternative, cooperative frameworks for solution.

By itself, the picture in Table 3.1 is non-instructive, the predicted behavior depends on the rewards, positive or negative, that the countries perceive for the different cells. Table 3.2a restructures the matrix

Country and Behavior

		A: Passive	A: Aggressive
B	Passive	X / X	0 / 2X
	Aggressive	2X / 0	−Y* / −Y*

Y* > X > 1

Country and Behavior

		A: Passive	A: Aggressive
B	Passive	1 / 1	0 / 2
	Aggressive	2 / 0	−Y / −Y

Y=Y*/X > 1

Table 3.2 (**a**) Costs and benefits (**b**) Normalizing on X

to allow for some analytic clarity, using the following simplifying assumptions:

- The potential benefits and costs are equal for the two countries. An alternative assumption is that the countries, for instance, see different costs to mutual aggression. This possibility figures in the discussion in the next chapter.
- The gains from some exclusive control of the maritime resources (2X) are twice the benefits (X) seen by each disputant if they passively coexisted: simultaneous passive behavior divides the potential gains in half. We are saying there is one "pie" to be either kept or divided.
- Passive behavior in the face of your counterpart's aggressive behavior results not only in zero benefits but also in zero costs.[2]
- Open conflict, resulting from simultaneous aggressive behavior, results in costs ($-Y^* < 0$). To set the tone of the argument, we further assume that these are larger than simply the failure to gain exclusive territorial rights ($Y^* > X$): conflict exacerbates the resource and control losses.

Table 3.2b normalizes this matrix by dividing through by X.

This structure is formally equivalent to the game of *chicken*. In the stylized picture of this game: "*Two hooligans with something to prove drive at each other on a narrow road. The first to swerve loses face among his peers. If neither swerves, however, a terminal fate plagues both*" (Shor 2006). In this game, each player has an incentive to behave aggressively, but the costs are large if both behave this way.[3] Thus, each country has an incentive to act aggressively, but would regret landing in the lower right-hand cell.[4] In these matrices, neither row dominates the other, and similarly, for the columns—both sides lack the incentive to stay solely within one form of behavior, either passive or aggressive, irrespective of the behavior of the other.

There are two cells which are Nash equilibria: if the players find themselves in either the upper right or lower left cell, there is no simple shift by either player that can improve their situation. Depending on how the game is played, one of these cells could be the outcome. Perhaps, for instance, by convincing your opponent that you will only act aggressively, no matter what, you could get an acceptance of the Nash equilibrium that favors you.[5] Indeed, it may be that the relatively recent island building and increase in military forces being stationed in the South China Sea are attempts to convince opponents that they will be presented only with aggressive behavior and that their rational choice will be to be passive. While this may be the way things play out in the future, in the past we have seen all countries varying their behavior between aggressive and passive.

Past behavior suggests that the set of incentives faced by each country encourages them to adopt a *mixed strategy*—to act passively sometimes, and sometimes aggressively. In assuming the absence of cooperative behavior there is a Nash equilibrium with mixed strategies—a paired set of behaviors that restricts each country's freedom of choice to the expected gains from aggressive behavior equaling that from the alternative behavior.[6] Using our notation, country A should set the probability of behaving passively (q) such that the expectation of gain for country B is equal, no matter which behavior B chooses, passive or aggressive. (Webster 2009, pp. 144–145) In Equation (Eq. 3.1) we set the expectation of gain for country B by behaving passively (left-hand side) equal to the expectation of gain to behaving aggressively (right-hand side):

$$q*1+(1-q)*0 = q*2+(1-q)*(-Y) \qquad \text{(Eq. 3.1)}$$

As our matrix is symmetrical with respect to the two countries, this expression is good for country B also. Solving this relationship, using our notation in Table 3.2b, the probabilities that each country would independently behave either in a passive fashion (q) or aggressive fashion (1-q) are given below:

$$q = Y/(1+Y) \qquad \text{(Eq. 3.2)}$$

$$1-q = 1/(1+Y) \qquad \text{(Eq. 3.3)}$$

This allows us to determine the explicit likelihood that either country would behave in a passive or aggressive fashion given some value for Y, the multiple of the costs of conflict to the possible gains from an exploitation of the resources. If the costs of real conflict were, for instance, 3 to 4 times the possible gains from cooperative behavior, we would expect the country to behave in a passive fashion 75–80 % of the time and act aggressively 20–25 % of the time. Figure 3.1 below shows how illustrative values for the costs of conflict as a multiple of possible gains would result in changes in behavior.

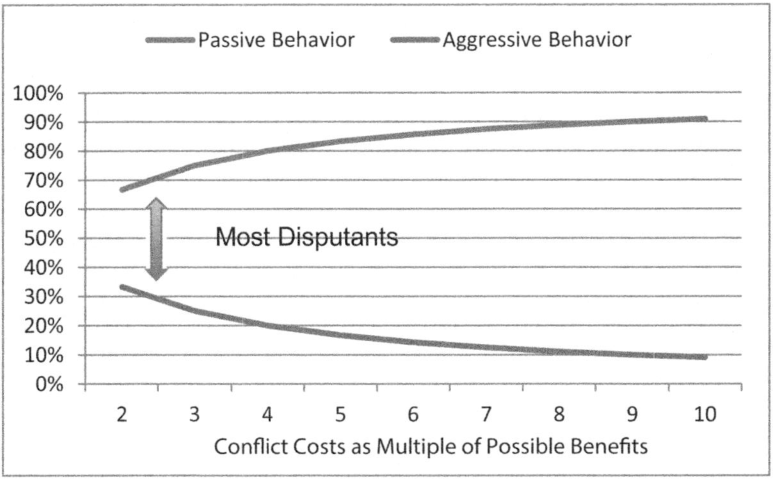

Fig. 3.1 Predicted behavior as a function of the perceived costs of conflict

One strong implication of this model is that, because each country acts independently of the other and each country has some positive probability of adopting aggressive behavior, there is some probability of the simultaneous selection of aggressive behavior or the probability of true bilateral conflict: (1–q)*(1–q). Again referring to Table 2.1, the bilateral pairings of disputant countries would suggest this event 1–14 % of the time.

We have seen clashes that have resulted in fatalities, especially between China and Viet Nam, but the most violent conflict between these two countries, in 1979, likely had little to do with competing claims to maritime rights in the South China Sea. (Stout 2014) This again illustrates the limitations of the simplistic model of game theory to explain the multifaceted behavior of real countries in real time in the real world. Governments do not approach every situation of interaction with others in the South China Sea independent of past behavior, in a purely bilateral fashion, and apart from other concerns. For instance, it may well be that China would not see the Philippines as acting as an individual player, but would see the USA as a possible partner to the Philippines in some circumstances. This element, the internationalization of the South China Sea disputes, is discussed below.

Game theory can help explain seemingly inconsistent behavior (aggressive behavior in one period versus passive behavior in another), but it can neither easily reflect the myriad of changing concerns that must be addressed in national policymaking nor the complexity of the decision-making processes. Decisions that affect the risk of conflict are made by different people at different levels of authority, by decision-makers with different individual interests. From the viewpoint of the center of national decision-making, it is likely that, while each country occasionally feels the need to behave in an aggressive fashion, they would moderate this behavior if it seems to be leading to true conflict.

The 2014 centennial of World War I was a stark reminder that this kind of calculation can go horribly wrong.[7] In the South China Sea, there is a particular risk that the ever-present interaction between fishing boats and coast guard or other maritime security vessels of the various countries can trigger a series of events that would end in real conflict.[8] O'Shea (2013, p. 4) warns of "the possibility of an accident involving military, coast guard, or activist vessels from both sides taking place in the disputed seas around the islands, which then spiralled out of control."[9] The ICG (2012, p. 20). expresses the concerns that: "The involvement of paramilitary vessels lowers the threshold for confrontation." Many of these pictures suggest how decision-making at different levels of government and by individuals with varying incentives and goals could lead to large, undesirable collective costs.

It is here that the island building, really a *militarization* of the South China Sea, poses real risks. This is especially so in the longer-term as larger numbers of naval vessels based in the area, with orders to protect their own nation's fishing vessels and police those of other countries, will invariably lead to increased armed confrontations. In the short run, as a facility is being built, while this act sends a clear aggressive message, it may do so, paradoxically, in a less confrontational manner than having a naval vessel actively harass a fishing vessel. Building up a submerged reef with dredged sand may allow nations to send a strong signal about their claims without risking contact that could spin out of control. This is clearly not always the case, and the initial occupation of an island has sometimes seen outright conflict, but this may be how the current Chinese and Vietnamese building is playing out.

It is over the long run that island and facility building is truly worrisome: base building reduces the administrative expense of aggressive behavior and likely enlarges the scope of responsibility for security forces. Island bases would allow, for instance, for a greater presence of maritime security vessels, both to undertake operations against other nations' vessels and to respond more quickly to similar activities by rival forces. A heightened military presence lends itself to wider use and the opportunities for real conflict between nations. Shear (2015, p. 3) suggests this is the official US Government position: "China's actions and increased presence could prompt other regional governments to respond by strengthening their military capabilities at their outposts, which would certainly increase the risk of accidents or miscalculations that could escalate." In the absence of mutually acknowledged territorial rights, it seems inescapable that confrontations will increase.

Figure 3.1 suggests that standing behind the current pattern of essentially risky behavior is a set of assumptions regarding the costs and benefits of engaging in these disputes: that the costs of aggressive behavior are not seen as large in relation to the possible gains.[10] This may have been true in the past, before China and Viet Nam were well integrated into the global economy, before the evolution and creation of *Factory Asia*. But for all countries, the costs and benefits of a particular policy at any particular time will change. Policy-making needs to be conducted against this framework of evolving national, regional, and global economic circumstances. A central thesis of this book, elaborated below, is that, over time the costs of aggressive behavior in the West Pacific Ocean are increasing, the net benefits of winning any violent confrontation decreasing. Today's East Asia bases its sterling economic performance on increasingly interdependent webs of trade and investment flows, but Factory Asia will come undone with international conflict.

In the world of game theory, this suggestion would imply that the probability of aggressive behavior should shrink and also the probability of conflict. In the real world, countries approach situations of confrontation from given positions—repeated games are not independent of past behavior. Policy positions, once hardened, are changed with reluctance. Referring again to the picture of the behavior of the disputants, the governments, as participating in a game of chicken, are saying "*But I can't back down now because / I pushed the other guys too far*" (Wilson and Christian 1964). This is also one reason why we have not seen collaborative strategies emerge in the South China Sea—without some change in the structure of the game, different behavior would take renunciation of past strategies.

In summary, the simple game theory model was used to (i) help explain why countries would alternate between aggressive and non-aggressive behavior in support of their claims to the South China Sea, (ii) draw attention to the need to more fully understand the costs and benefits of alternative scenarios, and (iii) motivate discussion about whether the conflict in the South China Sea can only be seen as non-cooperative or whether there are cooperative solutions to the problem. The next two chapters discuss the evolving costs and benefits of the conflict in the South China Sea. It will be argued that, in the western Pacific, the true costs from the current behavior are underestimated, the benefits exaggerated, and so the dangers of miscalculation, the risks of conflict, are large.

Notes

1. A few other researchers have used some aspects of game theory in examining behavior in the South China Sea. In a short piece, Tran (2009) talks through the situation as one of China *bargaining* with other Southeast Asian countries, emphasizing that China has more bargaining power and can dominate the other players unless they form a united front. Using a framework similar to the one used here, Tom Yam (2012), in a short newspaper column, explicitly analyzes the west Pacific Ocean conflicts using game theory, crucially making different assumptions about the nature of the costs to what we are calling mutually aggressive behavior (see footnote 4). Zhong (2013), in a wide-ranging master's thesis on the South China Sea, develops a three-player sequential game with China, the Philippines, and the USA as the players and military actions as options for play. The results of the game depend partly on the assumptions about outcomes of military clashes between these countries. Zhong concludes, "it is better for each party in this conflict to be calm" (p. 50) and that stability is a likely upshot.
2. An alternative assumption is that there are nonzero costs to passive behavior in the face of aggressive actions by an opponent. For instance, there may be the loss of transit rights. As long as these costs are less than those incurred when both parties act aggressively, the major conclusion of the chapter, that there is no dominant strategy for either side, holds. However, this assumption would affect the later discussion regarding the frequency of aggressive and passive behaviors.
3. Spaniel (2011) is a very clear YouTube video on this game.
4. Tom Yam (2012) assumes that mutually aggressive behavior (mutually "provocative" in the original piece) still results in positive gains to each country. This yields a Nash equilibrium in the lower right-hand cell. While clearly capturing some aspects of the real-life situation, that picture does not do justice to the actual costs of the conflict that might result from mutually aggressive behavior. This is the subject of Chap. 4.

5. Oskar Morgenstern, who along with John von Neumann developed game theory, emphasized that stability or equilibrium is very much dependent upon individual awareness or social processes and that this underlies some of the differences between the social and physical sciences. At a seminar by Morgenstern that I attended more than three decades ago, he made the point that a solid mass held in position by opposing springs could be said to be in a stable equilibrium. If one replaced the springs by two hot dogs and the central mass by a live dog, although the dog would be pulled equally in two directions, it would not be a stable situation. See Schotter (1990, pp. 107–108) for a fuller discussion of Morgenstern's views and his use of this picture.
6. There are other strategies that might be identified, for example, "tit for tat" where one country responds in some structured fashion to an earlier choice by the opponent. Similarly, we might try to characterize path-dependent games that would reflect "salami-slicing" or relatively small aggressive acts to win intermediate objectives in a longer campaign. See Haddick (2012) for a depiction in this manner of Chinese actions in the South China Sea. Cronin's (2014) description of "tailored coercion" is similar. These are useful insights; however, from the somewhat limited goals set for this chapter, there would be little benefit from exploring this within the framework of the game used in the text. The likely underlying instability of the policy framework and the decision-making process, the need to consider more than two players in any game, and the multiplicity of real-life goals discourage pushing this simple game theory framework too far in any one direction. The structure used, hopefully, is sufficient to support the hypotheses presented, especially that the disputants' behavior risks outright conflict.
7. From a different perspective, George Friedman (2014) expresses the concern: "Here I am more troubled … because nations frequently are not aware of what is about to happen, and they might react in ways that will surprise them." Friedman was commenting specifically on the relationship between Russia and Western nations, but it is perfectly applicable to the disputes in the western Pacific Ocean.
8. "Fishermen are both exploited by and exploit nationalist government sentiments and willfully push the boundaries of fisheries" Baker (2016).
9. Hayton (2014, pp. xiii–xv) also provides an all too believable scenario of how open conflict could result from a series of missteps. The Economist (2015) describes a delicate "dance" between Chinese and Japanese naval vessels in the East China Sea that can easily seem to risk triggering open conflict. Cronin (2013, p. 2), referring to the South China Sea disputes: "While these tensions appear manageable and are unlikely to trigger war (except through miscalculation or accident), they are trending in an unfavorable direction."

10. There are, of course, different interpretations of the observed behavior. A large CSIS study of US strategic interests in the Pacific concludes: "*China's tolerance for risk has exceeded most expectations*—as demonstrated by Beijing's increased operational tempo and construction of military airfields and facilities on seven features in the Spratly Islands" (2016, p. 5).

Bibliography

Baker, Rodger. 2016. Fish: The overlooked destabilizer in the South China Sea, 12 Feb. 2106. Stratfor. https://www.stratfor.com/analysis/fish-overlooked-destabilizer-south-china-sea. Accessed 21 Mar 2016.

Center for Strategic and International Studies (CSIS). 2016. Asia-Pacific rebalance 2025 capabilities, presence, and partnerships: An independent review of U.S. defense strategy in the Asia-Pacific. http://csis.org/files/publication/160119_Green_AsiaPacificRebalance2025_Web_0.pdf. Accessed 11 Mar 2016.

Cronin, Patrick M. 2013. The strategic significance of the South China Sea. For "Managing tensions in the South China Sea" conference held by the Center for Strategic & International Studies (CSIS) on 5–6 June 2013. http://csis.org/files/attachments/130606_Cronin. Accessed 23 Nov 2014.

——— 2014. The challenge of responding to maritime coercion. Washington, DC: Center for a New American Security, Sept 2014. http://www.cnas.org/Challenge-Responding-to-Maritime-Coercion#.Vu8M0xIrJDU. Accessed 20 Mar 2016.

The Economist. 2015. Small reefs, big problems. 25 Jul 2015. http://www.economist.com/news/asia/21659771-asian-coastguards-are-front-line-struggle-check-china-small-reefs-big-problems. Accessed 1 Aug 2015.

Friedman, George. 2014. Viewing Russia from the inside. Stratfor. http://www.stratfor.com/weekly/viewing-russia-inside#axzz3PDlLF1uL. Accessed 19 Jan 2015.

Haddick, Robert. 2012. Salami slicing in the South China Sea. *Foreign Policy*, 3 Aug. 2012. http://foreignpolicy.com/2012/08/03/salami-slicing-in-the-south-china-sea/. Accessed 23 Oct 2015.

Hayton, Bill. 2014. *The South China Sea: The struggle for power in Asia*. New Haven: Yale University Press.

International Crisis Group (ICG). 2012. Stirring up the South China Sea (II): Regional responses, report N 229, 24 July 2012. http://www.crisisgroup.org/~/media/Files/asia/north-east-asia/229-stirring-up-the-south-china-sea-ii-regional-responses. Accessed 22 May 2015.

O'Shea, Paul. 2013. Territorial disputes in Northeast Asia: A primer. Italian Institute for International Political Studies Analysis (ISPI) No. 182, June 2013. http://www.ispionline.it/sites/default/files/pubblicazioni/analysis_182_2013.pdf. Accessed 8 Dec 2014.

Schotter, Andrew. 1990. Oskar Morgenstern's contribution to the development of the theory of games. http://cess.nyu.edu/schotter/wp-content/uploads/2010/02/%E2%80%9COskar-Morgenstern%E2%80%99s-Contribution-to-the-Development-of-the-Theory-of-Games%E2%80%9D.pdf. Accessed 16 Feb 2016.

Shear, David. 2015. Statement of David Shear Assistant Secretary of Defense for Asian & Pacific Security Affairs before the Senate Committee on Foreign Relations, May 13, 2015. U.S. Senate Committee on Foreign Relations. http://www.foreign.senate.gov/imo/media/doc/051315_Shear_Testimony.pdf. Accessed 6 Mar 2016.

Shor, Mikhael. 2006. Game of chicken. Dictionary of Game Theory Terms. http://www.gametheory.net/dictionary/games/GameofChicken.html. Accessed 20 Feb 2016.

Spaniel, William. 2011. Game theory 101: Chicken. YouTube video. https://www.youtube.com/watch?v=sww-Zsl0IRY. Accessed 16 Feb 2016.

Stout, David. 2014. The last time China got into a fight with Vietnam, it was a disaster, 15 May 2014. TIME. http://time.com/100417/china-vietnam-sino-vietnamese-war-south-china-sea/. Accessed 17 Apr 2015.

Tran, Du. 2009. The South China Sea dispute: An unbalanced bargaining game. The Southeast Asian Sea Research Foundation. http://en.seasfoundation.org/index.php?option=com_content&view=article&id=48:the-south-china-sea-dispute-an-unbalanced-bargaining-game&catid=1:latest-news&Itemid=50. Accessed 16 Nov 2014.

Webster, Thomas J. 2009. *Introduction to game theory in business and economics.* Armonk: M.E. Sharpe.

Wilson, Brian, and Roger Christian. 1964. Don't worry baby. Attribution available. http://www.rollingstone.com/music/lists/the-500-greatest-songs-of-all-time-20110407/the-beach-boys-dont-worry-baby-20110526; lyrics available http://beachboyssongs.com/lyrics-page-1/lyrics-dont-worry-baby/. Accessed 6 Mar 2016.

Yam, Tom. 2012. Game theory and the South China Sea conflict. South China Morning Post, 16 Nov 2014. http://www.scmp.com/article/1006824/game-theory-and-south-china-sea-conflict. Accessed 16 Nov 2014.

Zhong, Xinhui. 2013. The gaming among China, the Philippines and the US in the South China Sea disputes. Master Thesis, Development and International Relations, Aalborg University, Denmark. June 2013. http://projekter.aau.dk/projekter/files/76994735/Thesis_final_edition.pdf. Accessed 16 Nov 2014.

4

The Economic Context: Costs and Vulnerability to Conflict

Abstract This chapter argues that the costs of the present disputes are likely larger than perceived by policymakers and are increasing. The chapter emphasizes the importance of the interwoven regional trade, including tourism flows. Cross-border investment is also crucial, especially given the evolving Chinese economy and its need for targets for outward foreign investment. The chapter concludes that these interrelationships—Factory Asia—cannot prosper with regional conflict. Recent experience also shows that these countries are vulnerable to macroeconomic shocks emanating in regional neighbors. Finally, as the ASEAN Economic Community evolves, a lack of resolution to the South China Sea disputes will likely prevent China from playing a useful role.

Keywords ASEAN • China • Trade • Foreign Direct Investment (FDI) • ASEAN Economic Community (AEC) • Southeast Asia

Real conflict over west Pacific territorial claims could be horribly costly. As noted earlier, international trade and cross-border investment in East and Southeast Asia have been the prime ingredients in the region's growth and poverty reduction. *Factory Asia* is the term used to refer to the basing of global value chains in China and Southeast Asia, providing assembly and intermediate manufacturing for world markets, fueled by cross-border investment, especially from South Korea and Japan, and facilitated by

D.J. Green, *The Third Option for the South China Sea*,
DOI 10.1007/978-3-319-40274-1_4

technological changes that lowered oceanic shipping costs.[1] This world-shaping process rested on international trade and cross-border investment and was, without conscious overall direction, a collaborative effort among the firms and governments in East and Southeast Asia.

Factory Asia cannot survive real conflict in its present form either in the East or the South China Sea. While we have not reached this point, the ongoing tensions between China and Japan, the Philippines, and Viet Nam, respectively, have led to economic impacts foreshadowing exactly this result. As shown in Table 4.1, after a confrontation, perhaps between

Table 4.1 Economic costs attributed to maritime tensions in the Western Pacific

Parties	*Timing and reported impacts and costs*
China–Japan Conflicting claims to Senkaku Islands (Japan)/ Diaoyu Islands (China)	2010 after Japanese detention of Chinese fishing boat: Restrictions on shipments of rare earth minerals from China to Japan. (Bradsher 2010) 2012 after the Japanese Government assumed control of disputed islands: Reports that "Chinese protesters smashed Japanese cars," Chinese consumers voiced a reluctance to buy Japanese cars, and there was a drop-off in sales in China of Japanese cars (Bhattacharya 2014; See also Roberts 2012). "[M]ajor Japanese businesses temporarily closed hundreds of stores and factories across China. Honda closed 104 outlets and received 2120 canceled orders, while Nissan, Toyota, and Mazda factories also suspended business" *SupplyChain* Magazine (2012). Bilateral investment into China from Japan and vice versa falling off "by almost 50 percent in the first half of 2014 compared to the same period a year ago. …[and] by 23.5 percent in 2013"respectively. (Domínguez 2014).
China–Philippines Conflicting claims to the Scarborough Shoal (Huangyan Island in China) and some of the Spratly Islands	After May 2012 standoff at the Scarborough Shoal: Imports of fruit from the Philippines held up in quarantine and Scheduled airline flights from China to the Philippines canceled disrupting tourist travel (De Castro 2013, p. 7).
China–Viet Nam Conflicting claims to islands in the Paracels and the Spratlys	After May 2014 deployment of a Chinese oil drilling rig in waters claimed by Viet Nam near the Paracel Islands and violent confrontations between Chinese and Vietnamese vessels (The Economist 2014b): Rioting in Viet Nam resulted in deaths and extensive looting and destruction of Chinese- and Taiwanese-owned factories Vu and Hsu (2014).

naval vessels or between naval vessels and fishing boats, there has sometimes been a reaction in one of the countries that affects the broader economies. After maritime clashes, for instance, Philippine exporters of bananas saw their goods held up by Chinese officials, Chinese protestors destroyed Japanese cars, and Vietnamese protestors damaged Chinese and Taiwanese factories. In some instances, lives were lost and in many instances trade was disrupted.

In and of themselves, these could be isolated events that have little or no lasting impact but fit in an atmosphere of continuing maritime confrontation they lead, over time, to poisoning economic and political relationships, heightening the potential for real conflict. The above assumption that the countries could act independently in a sequence of repetitive games does not reflect well the political economy of East and Southeast Asia. Equally important, while to-date, the overt economic impacts have not been large enough to disrupt regional economies; increasingly the past experience is not reflective of how costly even a short-term, open conflict could be. This is because of several interrelated trends: (i) the increasing trade intensity of the economies, (ii) the importance of and changing nature of regional cross-border investment, (iii) the utility of smooth political relations as the ASEAN Economic Community takes shape, and (iv) how continued market liberalization heightens economic vulnerability to shocks. These arguments will be developed below. In addition, this Chapter will discuss how the growing importance of the ASEAN Economic Community suggests that these conflicts cannot solely be seen as a series of bilateral problems, but rather continued tensions will naturally become multilateral in nature.

4.1 Trade Intensity: Trade Dependency

It is difficult to overstate the role of trade in the economies of East and Southeast Asia and that of the South China Sea as a conduit for regional and global trade. With respect to the latter, the U.S. Energy Information Administration estimates that "More than half of the world's annual merchant fleet tonnage passes through the" narrow straits at the southern part of the South China Sea with much of that moving through to the major international ports of the region. (US EIA 2013)

Figure 4.1 shows the increasing trade intensity of the countries most involved in the west Pacific trade disputes, looking at the ratio of total exports and imports of both goods and services to gross domestic product. The initial figures in this section visually distinguish between the two different regions of the west Pacific Ocean. Thus, the data for Southeast Asia are to the left in Fig. 4.1, and Northeast Asia to the right. This chapter will not generally report on the situation of Taiwan. As noted earlier, Taiwan's claims generally "mirror" those of China. More importantly, Taiwan's activities in the South China Sea are very much secondary to the relationship between the governments in Taipei and Beijing. Actions by the Taiwanese government are likely to be constrained by a need not to destabilize relations with China and are unlikely to figure prominently in events in the South China Sea. Finally, data on the Taiwanese economy are not always reported in a manner comparable to other economies.[2]

For many of the countries in the figure, the economies have become more trade intensive over the three decades, sometimes dramatically so. In particular, from the 1980s to the 2000s, the trade intensity of the Chinese economy has increased more than 60 %, raising the stakes for the country should there be a major disruption in international trade.

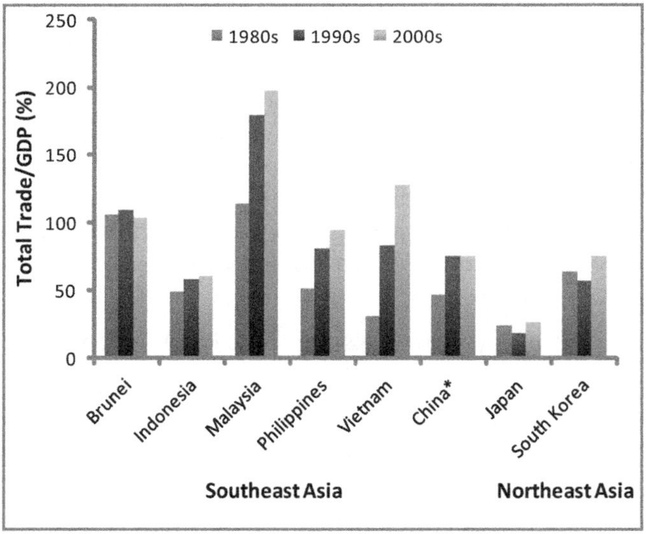

Fig. 4.1 Total trade/GDP.

Notes: Decade averages of the sum of exports and imports of both goods and services over GDP. World Bank, World Development Indicators. Values for the 1980s were not available for all years. China* includes Hong Kong and Macau with the combined data aggregated from the individual economy reports weighted by the respective share of the reported US dollar value of overall GDP

Rising trade intensity has underwritten substantial increases in the standard of living. China's recorded per capita GDP grew from $221 in 1980 to $3863 in 2014 (constant 2005 US dollars).[3] Smaller, but still impressive, increases were seen in most of the countries involved in these maritime disputes, especially South Korea and Viet Nam. This relationship between international trade and economic growth is not lost on policy-makers—this is a region that believes in trade, as an instrument to encourage growth and development. (Green 2008b)

Much of the growth-fueling international trade in East and Southeast Asia is regional in nature, especially involving the Chinese economy. Figure 4.2 illustrates this through the use of a logarithmic scale to show the exploding value of bilateral trade between China (and Hong Kong) and the other countries that dispute China's west Pacific claims.[4] Except for relatively small Brunei, the total trade flows (exports plus imports) from Southeast Asia to China have grown to some tens of billions of US dollars and an order of magnitude larger for Japan and South Korea. This, of course, reflects one aspect of *Factory Asia*—the increasing use of distributed production sites linked together by international trade.

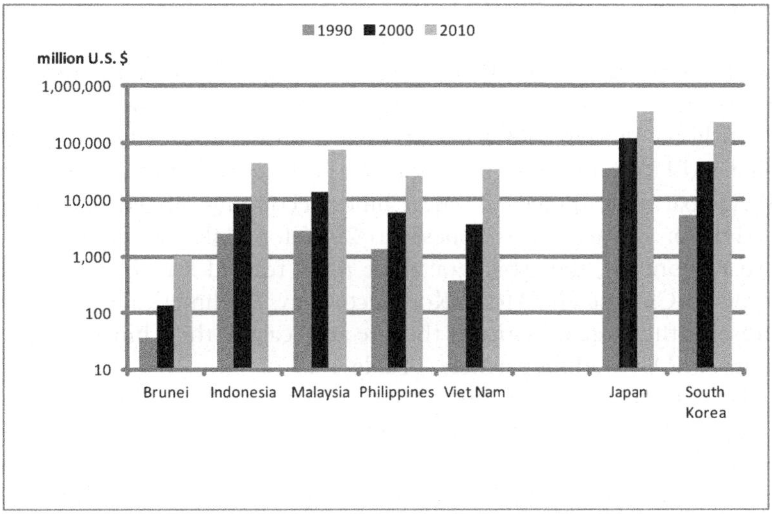

Fig. 4.2 Total trade with China and Hong Kong.
Notes: Asia Regional Integration Center website citing International Monetary Fund Directions of Trade. Values are the average of the reports from Hong Kong and China and the trading partner. (Data for Macau were not available)

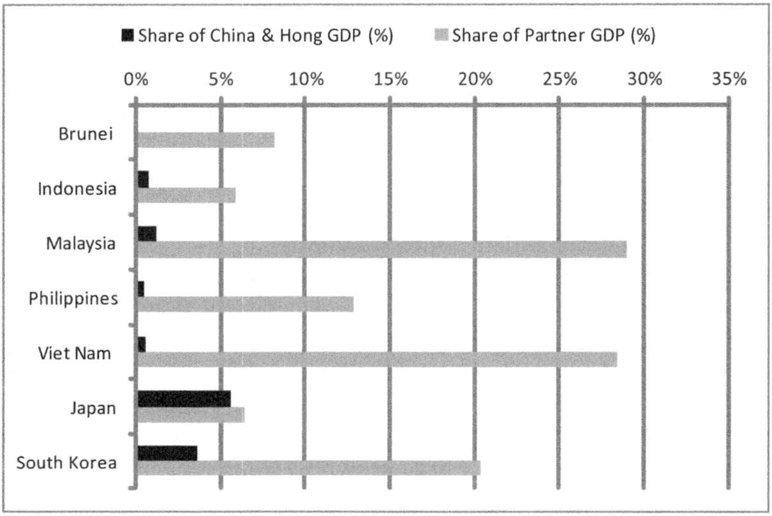

Fig. 4.3 2010 GDP share of total bilateral trade.
Notes: For trade flows, Asia Regional Integration Center website; GDP data from Trading Economies website

The interdependency of this group of nations, and by extension the costs of trade-stopping conflict, is further put in perspective by Fig. 4.3 which shows the bilateral trade flows as a proportion of GDP of either China and Hong Kong, or the respective ASEAN trading partner. For the ASEAN partners, trade with the Chinese economy ranged from 6–8 % of GDP for Brunei and Indonesia to 29 % for Malaysia and Viet Nam. In sharp contrast, these bilateral trade flows totaled less than 2 % of the combined Chinese and Hong Kong economy. Arithmetically, this simply represents the huge disparity in the size and scale of the Chinese economy and its ASEAN trading partners. (Table 4.2)

These figures suggest huge, immediate costs to China's ASEAN trading partners if disputes shut down trade and much smaller costs to China. China's ASEAN trading partners would clearly face serious economic costs from any trade fall-off if the present level of confrontation turns more violent.[5] It is this picture that lends itself to the impression that the ASEAN trading partners "have more to lose than" China if tensions become more pronounced.[6] This is not the case if confrontations

Table 4.2 2010 China GDP/ASEAN trading partner GDP

Brunei	506.7
Indonesia	8.3
Malaysia	24.6
Philippines	31.4
Viet Nam	54.1

Notes: http://www.tradingeconomics.com
Accessed April 17, 2016

escalate between China and Japan or between China and South Korea, or if regional trade generally is disrupted.[7] The risk to regional trade is a real one. The Asian Development Bank puts the issue very succinctly: "Because most intra-ASEAN activities are through the production network, shocks emanating from one ASEAN country can disrupt the region's supply chain and, hence, destabilize output in other members" (ADB 2013a, pp. 15–16).

Moreover, the notion that any country could easily ride out regional conflict is based on a very simplistic and inaccurate picture of the nature of a modern, market-based economy. Disruption in international trade could have a wide range of impacts on growth through myriad channels, one of which is tourism (discussed in Box 4.1 and Appendix A); another is foreign-financed investment flows, discussed below.[8]

Box 4.1 Tourism Is Vulnerable to West Pacific Conflicts Conflict in the west Pacific Ocean that deters the ability to travel, transport goods, or transit between countries and regions would have a tremendous impact on merchandise trade, but would also affect tourism–part of the important and growing services trade in East and Southeast Asia. Tourism plays a large role in many of the countries in this report. Even in the huge Chinese economy, earnings from tourism are equivalent to one of the larger export industries. It is a sector that, looking forward, presents tremendous potential for East and Southeast Asian economies.

However, this is a potential that can be crippled by political disputes. Worsening relations between Japan and South Korea in 2008 resulted in a sharp reduction in Korean travelers. Similarly, Chinese travel to Japan in mid-2012 slumped with heightened political tension between the two countries. The fall-off in travel in both cases could be compared to that witnessed in the wake of the 2011 tragic earthquake and tsunami. In the South China Sea interruptions to travel were seen in the bilateral flows between China and Viet Nam in 2014/15, and in 2012, in the number of Chinese visitors to the Philippines. (Appendix A discusses this in more detail.)

4.2 Foreign Direct Investment: Ties That Bind

In the late 1970s, beginning an era of reform, China turned away from being an autarchic, command economy to relying increasingly upon market-based institutions, emerging as a lynchpin of global trade. One important part of these reforms included opening the economy for foreign direct investment (FDI). The impact of FDI continues to this day. "According to the Ministry of Commerce (MOFCOM), foreign invested enterprises account for over half of China's exports and imports; they provide for 30 % of Chinese industrial output, and generate 22 % of industrial profits while employing only 10 % of labor—because of their high productivity" (World Bank 2010).

Concurrent with the opening of China to international markets, the Plaza Accord in 1985 encouraged Japanese firms to move intermediate production lines to other Asian economies (Thorbecke and Salike 2011). This accelerated with China's membership in the World Trade Organization in 2001. Japanese FDI and, somewhat later, South Korean FDI, were important elements in the transformation of the Chinese economy as well as in those ASEAN economies actively involved in the South China Sea maritime disputes. Figure 4.4 shows that, despite the deepening of the Chinese economy, FDI from East and Southeast Asia still account for more than one-quarter of the stock of FDI in China.[9] Japan and South Korea remain important financiers of the Chinese economy.

These investment flows, providing the framework for Factory Asia, will not survive untouched by open conflict. If even a few borders close, the nature of these relationships will be reassessed with businesses reducing investments that have hitherto relied upon open regionalism. As noted above in Table 4.1, we have already seen disruptions to operations of foreign-owned factories in China and Viet Nam in response to the still

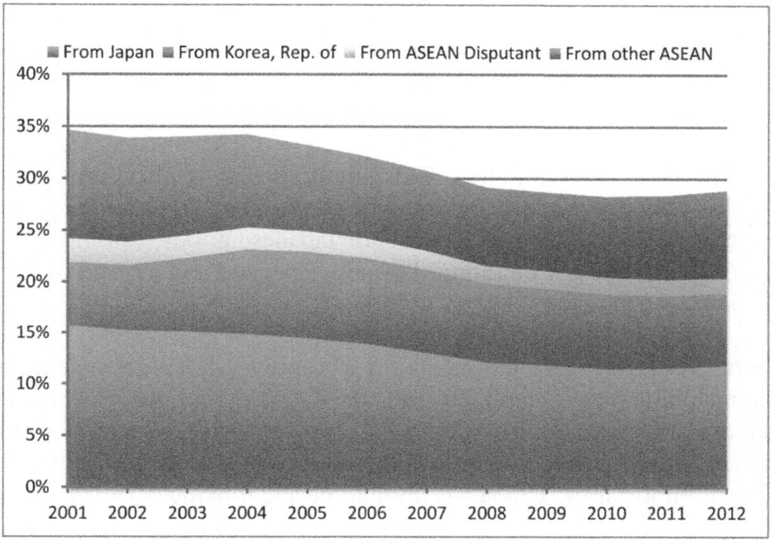

Fig. 4.4 Share of foreign direct investment in China* by origin.
Notes: Data from UNCTAD FDI/TNC database (http://unctad.org/en/Pages/DIAE/FDI%20Statistics/FDI-Statistics-Bilateral.aspx). China*: The data are relative to total stock measures in China minus that accounted for by Hong Kong and Macau. Data were not available for all countries. ASEAN disputants include Indonesia, Malaysia, Philippines, and Vietnam. Other ASEAN includes Singapore and Thailand.

relatively low-level conflicts in the South and East China Seas. One or two instances of violence may not factor heavily in business planning. But if tensions heighten, should consumer protests and boycotts violent or otherwise become more common, undoubtedly every firm would review its plans, likely reducing bilateral cross-border investment and business presence.

There are considerable anecdotal reports of Japanese businesses diverting investment plans from China to Southeast Asia at least partly in response to west Pacific conflicts.

- Hayashi and Negishi (2013) conclude that Japanese businesses are shifting focus away from China, partly due to maritime confrontations: "They [businesses] point to mob violence a year ago against Japanese businesses in China, as tensions rose over a group of uninhabited islands in the East China Sea."

- Bradsher (2014) notes some multinational firms are moving resources from China to Southeast Asia for a wide range of reasons including that "China's diplomatic and trade ties to Southeast Asia have been strained by its increasingly assertive claims to control over practically all of the South China Sea."
- The Asian Development Bank identifies the connection between trade and foreign investment decisions and political tensions, noting that there was "a PRC boycott of Japanese goods, causing a sharp drop in trade and engulfing Japanese investment in a cloud of uncertainty" (2013a, p. 28).

While the growing Chinese economy is progressively less dependent upon FDI, including from trading partners, any reduction in ongoing operations (including the maintenance of capital) by foreign-owned enterprises would further exacerbate the growth slowdown being experienced in China.

More importantly, China's economic needs, looking forward, are not the same as in the last few decades and this has to be taken into account in correctly understanding the possible costs of conflict in the South China Sea. In particular the role of FDI in the Chinese economy is changing. FDI was an important stimulus for the development of Factory Asia and China's unprecedented growth in the past two decades. But as the Chinese economy grows, the importance of *inward* FDI has diminished and will continue to do so, falling from 15 % of gross fixed capital formation in 1995 to 3.1 % in 2012.[10] Concurrently, as Fig. 4.5 shows, with the evolution of the Chinese economy, *inward* FDI is being matched by *outward* FDI. To-date much of the outward investment flows have been to (i) develop natural resource supplies, reflecting the huge rise of living standards and demand for raw materials for processing; and (ii) buying iconic properties in Western economies.[11] This is changing. In the near future, Chinese outward FDI should shift to extending value chain production abroad and a natural focus would be Southeast Asia.

Southeast Asia generally has become an important target for outward FDI from China. (Devonshire-Ellis 2014) While some of this continues the search for natural resources, increasingly it reflects the changing nature of the Chinese economy. Chinese firms are responding to the increasing domestic wage levels and investing in their near neighborhood where labor costs are lower. Devonshire-Ellis (2014) has estimated that average wage costs in China are roughly 2–3 times that of Indonesia and the Philippines, with wage costs in the principal manufacturing areas even higher. Chinese outward FDI

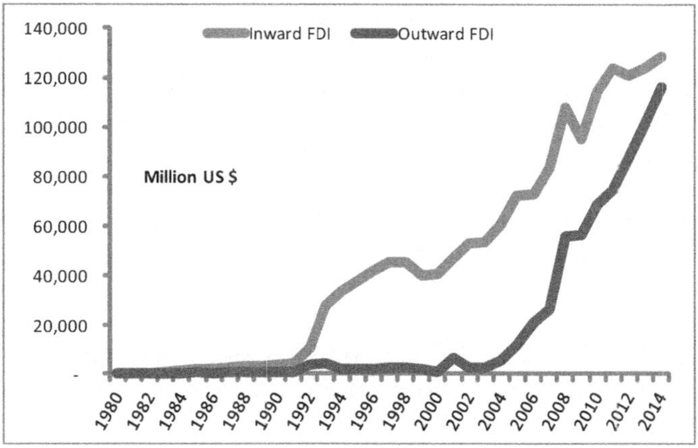

Fig. 4.5 Chinese FDI, inward and outward.
Notes: UNCTAD statistics. These flows do not include those to or from Hong Kong and Macau

is at an early stage; Hong (2013, p. 2), for instance, finds that China's FDI "in Vietnam is concentrated on low-skill, labour-intensive manufacturing and tends to be relatively small-scale." However, absent political disruption, with continued trade-facilitating policies and investments, Chinese firms would increasingly move to Southeast Asia, complementing a domestic economy increasingly in need of options to counter softening growth.

As discussed below, the movement, albeit in an uneven fashion, toward a single market in Southeast Asia, under the institutional structure of the ASEAN Economic Community (AEC), should enhance the opportunities for outward Chinese FDI. Continued lack of resolution to the maritime disputes will be a significant inhibition to realizing this potential. The lack of a peaceful resolution to South China Sea disputes could be a serious stumbling block to what otherwise would be a natural extension and deepening of Southeast Asian–Chinese economic relations.

4.3 Bilateralism Versus Multilateralism

At this point, it is worth re-examining one of the crucial assumptions used in the sections above: bilateralism, that one can model the South China Sea conflicts on a two-country basis. This was our initial starting point; however, there are real questions as to whether bilateral disputes in the

South China Sea, especially those involving China and an ASEAN member state, would remain purely bilateral in nature if real conflict broke out, or whether there would be regional consequences. This section of the book discusses the potential role of ASEAN, and a later chapter, the possible involvement of non-regional players such as the USA.

Regional consequences stem from the current nature of the South China Sea conflicts as generally involving China on one side, and on the other, one-half of the membership of ASEAN. While we have seen open conflicts between ASEAN members, the current heat in the conflict is centered on China and some ASEAN members. ASEAN as an organization has been involved in these disputes since the mid-1990s.[12] This involvement has not only echoed specific instances of confrontations, but has also reflected the changing nature of ASEAN as an institution since the founding in 1967. ASEAN of the 1970s and 1980s was much smaller than today, comprising Thailand and the archipelagic Southeast Asian countries (Brunei, Indonesia, Malaysia, the Philippines, and Singapore). Only subsequently did the remaining mainland countries (Cambodia, Lao PDR, Myanmar, and Viet Nam) join. Concomitantly to the change in membership, the focus of the organization started to evolve, with ASEAN increasingly seen as centered on economic cooperation through the creation of an ASEAN Economic Community. In spite of this, it maintains institutional arrangements that recall its inception as "a political alliance to limit the spread of communism in Southeast Asia" (ASEAN UP 2014).

Currently ASEAN is organized on three pillars, the (i) ASEAN Political-Security Community (APSC), (ii) AEC, and (iii) ASEAN Socio-Cultural Community. Under the APSC, while not echoing anything like the collective security arrangements of North Atlantic Treaty Organization (NATO), the members have pledged to "regard their security as fundamentally linked to one another" (ASEAN, APSC). The institutional arrangements involve complex regular consultations not only at the ministerial level among the ASEAN members, but also with non-members, including China, in the ASEAN Regional Forum.

Given this level of dialogue on security issues it would be strange indeed if ASEAN had not attempted to deal with South China Sea issues. Efforts have included the negotiation over and signing of a Declaration on the Conduct of Parties in the South China Sea in 2002 between all members of ASEAN and China (ASEAN DOC 2002). Among other things, the Declaration commits to non-violent dispute resolution: "The Parties concerned undertake to resolve their territorial and jurisdictional disputes by peaceful means, without resorting to the threat or use of force, through friendly consultations and negotiations by sovereign states directly concerned." In spite of the clear promise, implementing arrangements were never finalized (Thayer 2012). Table 1.1 and Fig. 2.1 show that the commitment to eschew violence (and certainly the threat of violence) has not been adhered to. The ineffectiveness of the Declaration on the Conduct of Parties reverberated in the failure of ASEAN to issue a joint statement at the conclusion of the ASEAN Ministerial Meeting in 2012, the first time the organization had not found consensus on this document (BBC 2012).

In part, this diplomatic failure simply reflects the wide divergence of interests among the various ASEAN members with respect to the South China Sea and with China. Some of the disputes, especially China–Philippines and China–Viet Nam, have occasioned considerable violence in the past two decades; others, involving Brunei, Indonesia, and Malaysia, have seen less open conflict. The remaining ASEAN countries, Cambodia, land-locked Lao PDR, Myanmar, Singapore and Thailand

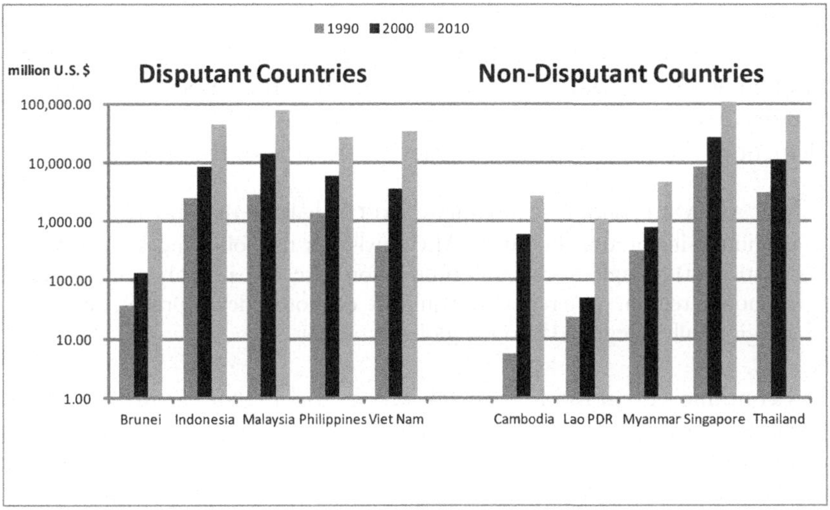

Fig. 4.6 Total trade ASEAN and China and Hong Kong.
Notes: Asia Regional Integration Center website citing International Monetary Fund Directions of Trade. Values are the average of the reports from Hong Kong and China and from the trading partner. [Data for Macau were not available]

have no comparable claims in the South China Sea. These nations, however, have considerable economic ties to China. Figure 4.6 reproduces some of the data from the earlier Fig. 4.2, showing the total bilateral trade between Southeast Asia and China (and Hong Kong). This chart swaps out the earlier data for Northeast Asian countries and embeds data for those ASEAN members that have no immediate stake in the South China Sea conflicts. Individually and as a group, the non-South China Sea disputants have tremendous interests in maintaining and growing trade with China, interests that will keep them from sticking their oars into this troubled sea.

This basic impasse within Southeast Asia is unlikely to change, certainly not in the sense of ASEAN as an organization siding overtly with one of its members against China over the South China Sea. While ASEAN is unlikely to take sides in any South China Sea dispute under the APSC, if open conflict occurs it will affect the relationship between China and ASEAN, including the emerging ASEAN Economic Community.

4.4 The Potential of the ASEAN Economic Community

The earlier discussion established the importance to the Chinese economy of open access to ASEAN markets to continue the smooth operation of Factory Asia as well as to provide scope for outward FDI. An important element in this is the ongoing, if uneven, creation of the AEC:

> The ASEAN Economic Community (AEC) shall be the goal of regional economic integration by 2015. AEC envisages the following key characteristics: (a) a single market and production base, (b) a highly competitive economic region, (c) a region of equitable economic development, and (d) a region fully integrated into the global economy.[13]

Economic integration provides scope for higher growth through broader markets and economies of scale of production and more efficient production enforced through heightened competition for regional markets. Over the long-term, a single market for production and sales in Southeast Asia would likely enhance the draw for China both as a trading partner and for outward FDI.

Less obvious but equally important, peaceful resolution to the conflicts in the South China Sea would allow China to become more of a player in the rule-making or standard-setting that underpins economic integration. Setting common technical standards for the production and sale of goods and services, for reducing non-tariff trade barriers including for customs, immigration, quarantine, and safety (CIQS) is an enervating bureaucratic activity. But it is crucial to the eventual realization of the goal of improvements in productivity through economies of scale that could come from continued opening of regional borders. Work by McKinsey & Co. suggests that in Southeast Asia, looking across a wide range of products, the present fragmentation of markets and non-uniform standards can add 10–15 % to costs.[14] As ASEAN writes the rules for the AEC there can be real advantages for Chinese companies if the Chinese government is involved in this process.[15]

China is already a dialogue partner in many of ASEAN activities.[16] But it is hard to see this continuing to the degree necessary for China to be involved in rulemaking if there is a real rupture in diplomatic relations between China and any one of the ASEAN members. ASEAN works through consensus and sharp objections by any ASEAN member could likely bar China from formally being involved in the process (a seemingly endless round of meetings) to the extent required to help fashion regionwide standards.

4.5 Vulnerability to Macro Shocks

The discussion above has emphasized the costs to the South China Sea disputants of lost trade and investment opportunities should the conflicts escalate; of a failure to smoothly continue economic integration based on the ASEAN Economic Community. This picture of potentially large costs to both ASEAN and China still misses an important risk: conflict over maritime rights could act as a general shock to an economy, for instance, triggering capital outflows and asset market collapses. This is particularly true for Asian economies that have used trade as the spark plug for growth. ASEAN economies experienced this vulnerability in the 1997/98 Asian Financial Crisis and in 2007/08 with the Global Financial Crisis. Especially in the former, a negative shock in one country, spread like a contagious disease with asset and financial markets collapsing and trade spiraling downward. More recently, in 2013, many Asian economies suffered *taper shock* as asset markets reacted to suggestions from then-Federal Reserve Chair Ben Bernanke that the US central bank might soon tighten monetary policy (ADB 2013b).

While many East and Southeast Asian economies have experienced this vulnerability, it is new for China and Viet Nam. As these countries come to rely more on market-based institutions, they become more vulnerable to external macroeconomic shocks, something that could result from even a small interruption to international trade or capital flows. In a real sense, it is the success of the Chinese and Vietnamese governments in economic transformation that makes them vulnerable to the very generic problem of instability of the market economy. "This is the double-edged sword of relying on external markets for development: it provides a tremendous boost to the efficiency of the economy and long-term growth, but increases short-term vulnerability to external shocks" (Green 2010, p. 7).

This is particularly important for the Chinese economy. Over the last few decades, as the Chinese economy has evolved, capital flows in and out of the country have been progressively liberalized. In the long run, freeing international capital flows from bureaucratic control provides for an economy that can more efficiently use economic resources, but it brings with it short-run vulnerabilities—the possibility that capital can leave in volumes sufficient enough to effect contemporary conditions.[17]

Economic shocks can arise from many different sources, including conflict and the reaction to aggressive behavior in pursuit of maritime claims. The International Monetary Fund (IMF) Deputy Managing Director, Naoyuki Shinohara identified "Rising territorial conflicts in the waters in Southeast Asia" as among the serious risks to some countries and to the region's potential for stable growth. The senior IMF official was further quoted as saying, "If the tension gets higher, it could have large impacts on the economies, especially in the case of this region where the countries are interconnected through supply chains" (Francisco 2014).

The Fund's concerns are more fully noted in the Article IV consultation reports. These documents, generally done on an annual basis provide a comprehensive picture of a country's economy, including sometimes a summary of staff views on economic vulnerability. In the case of the countries discussed in this book, in both Northeast and Southeast Asia, there have been a number of notices in which IMF staff refer to the possible consequences of "geopolitical" incidents having an impact on the macro economy. Some of these notices are gathered together in Table 4.3. The reports do not always provide specific detail to confirm that "geopolitical" events include conflict in the West Pacific Ocean; however, it seems likely that this would be one worry and the above statement by a senior official of the IMF would seem to support this interpretation. Chapter 2 shows that those countries which the IMF staff identify as being at risk from

Table 4.3 Consolidated risk assessment matrix IMF Article IV consultations

Country	*Likelihood of Occurrence*	*Impact on Economy*	*Description*
China	High	Medium	"Surges in global financial market volatility, triggered by geopolitical tensions or revised market expectations on UMP exit in the United States or emerging market fundamentals" IMF (2014c, p. 68).
Japan	Low	Medium	"Disruptions triggered by geopolitical incidents in East Asia.... An incident in the East China Sea could lead to disruptions in financial flows, and possible trade routes with a direct impact on global market sentiment and a flight to safety" IMF (2014d, p. 42).
Philippines	–	–	"A sharp slowdown in China or emerging markets more broadly, or a major geopolitical incident that disrupts regional or global trade, as well as commodity and financial flows, would also adversely impact the Philippines" IMF (2014e, p. 6).
Viet Nam	High	Medium	"Disruptions triggered by geopolitical incidents in East Asia" IMF (2014f, p. 9).
S. Korea, Malaysia, Indonesia	No mention of risks associated with maritime disputes. IMF (2014a, 2015a, b)		
Brunei	Only a press release or statistical note is generally released. See, for example, IMF (2014b, 2015c).		

Note: *UMP* Unconventional Monetary Policies

geopolitical events are those that have been behaving more aggressively with respect to their claims.

A common element in macroeconomic vulnerability is the potential for capital flight, for the sudden movement of financial resources out of a country. All Asian governments are well aware that capital flows are capricious and have endeavored to guard against the impact of capital flight while maintaining an economy that is open to financial flows supporting trade and cross-border investment. Figure 4.7 shows that foreign exchange reserves under the control of monetary authorities have increased both absolutely and in relation to the level of trade (as measured by imports). Thus, in the event of a precipitous degree of capital flight, the

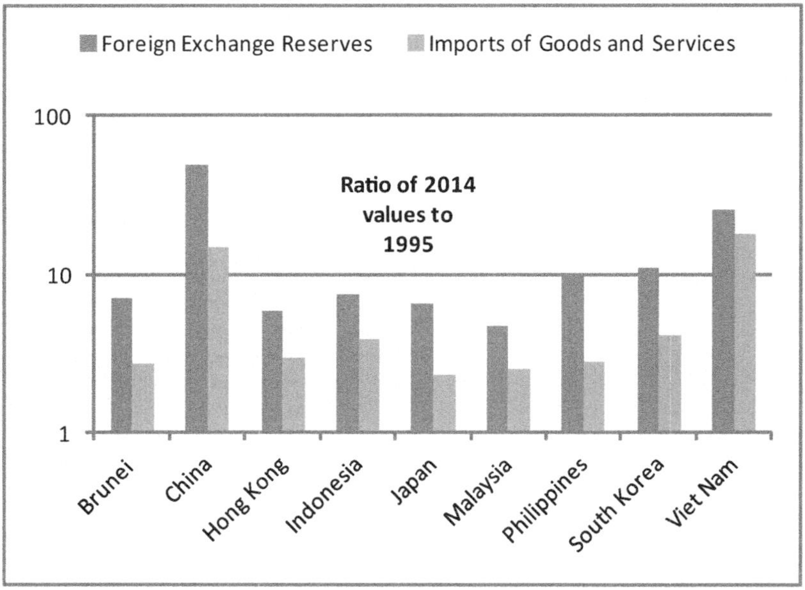

Fig. 4.7 Imports and foreign exchange reserves.
Notes: Data source, World Bank web site, World Development Indicators. Foreign exchange reserves are those under the control of monetary authorities including gold. For Brunei the time period was 1999–2014. The vertical scale is logarithmic

Asian maritime disputants today have a greater ability to finance imports than has been true in the past. This, however, is not the same as being able to insulate the country's economy from the impact of conflict in the western Pacific Ocean. Capital flight, precipitated by maritime conflict, for instance, could have disastrous impacts on asset markets and investment decisions throughout the region—impacts difficult for monetary authorities to mitigate even with large foreign currency reserves.

The experience of the unpredictability of shocks and the degree to which shocks can pass from one country to another (economic contagion) has led to an understanding that there is a need to face potential problems cooperatively. This has led to the creation first of the Chiang Mai Initiative (CMI) and the follow-up Chiang Mai Initiative Multilateralisation (CMIM). These efforts bring together the 10 ASEAN members plus the three large Northeast Asian countries: China, Japan, and South Korea to provide "regional foreign exchange liquidity support" (Sussangkarn 2012). The intent is to provide relatively unconditional foreign exchange to mitigate pressure on exchange rates in moments of crisis. The mechanism relies heavily upon provision of reserves by the larger economies in Northeast Asia to countries in need.

There is every reason to believe that this mechanism might work, at least to some extent, to mitigate the impact of a crisis from beyond the region—a sudden change in monetary policy in the USA or Europe for instance. It is counterintuitive, however, to presume that the institutional arrangements could be relied upon if the precipitating shock was the result of conflict between two of the member countries. Indeed it is likely that continued failure to peacefully resolve west Pacific maritime claims reduces the public confidence that cooperative mechanisms such as CMIM would function properly.

Notes

1. "Currently, *Factory Asia* maps a vast regional production network with the Republic of Korea and Japan as major outsourcing countries, and the PRC and most South-East Asian economies as assemblers of parts and components into final products." Choi and Rhee (2014, p. xii) As a microcosm of Factory Asia, WTO (2011, p. 16) provides a schematic picture of the interrelated automotive parts industry in ASEAN. Many factors have contributed to this system; for instance, the same publication notes the importance of the container revolution (p. 30).
2. See, for example, the explanatory note by the World Bank, available at https://datahelpdesk.worldbank.org/knowledgebase/articles/114933-where-are-your-data-on-taiwan [accessed 5 Jul. 2016].
3. Data from World Bank, World Development Indicators available http://databank.worldbank.org/data/views/variableSelection/selectvariables.aspx?source=world-development-indicators [accessed 23 Apr. 2016; data series for Viet Nam begins in 1984].
4. Well-known issues cause trading partners to report different flow data. The table uses the average of the reports from Hong Kong and China and from the trading partner. In all cases, the reports from China and Hong Kong are larger than the trade reported for any of the seven partners; as much as 74 % in the case of the Philippines. Data for Macau were not available.
5. The Asian Development Bank provided a different, but similarly telling perspective on the growing interrelationship between ASEAN economies and China: how lowered GDP growth in China would affect its Southeast Asian trading partners. "Absent any other distortion, GDP being 1 percentage point less than baseline in the PRC tends to reduce ASEAN's output by almost 1/3 of a percent from its baseline in the same year." The impact on the Malaysian economy was the largest identified for any of the ASEAN members (ADB, 2013a, p. 12).

6. De Castro (2013, p. 7) quoting the president of the Philippine Exporters Confederation (PHILEXPORT), Sergio Ortiz. De Castro also provides other, similar observations. Zhong (2013, especially p. 49) also works with the assumption that the Philippines would lose more in conflict with China.
7. Frost (2008, p. 166) makes the point that China is very dependent upon Asian countries as a group, especially for imports, as well as for meeting its need for foreign-financed investment.
8. *SupplyChain* (2012) notes some observers argue whether Japan or China would be affected more from conflict. The discussion in the text tries to show these arguments cannot be very convincing.
9. Economic statistics, including those related to the ownership of the capital stock in China by foreigners, are complicated by the separate reporting of data for China and Hong Kong (and Macau). For simplicity, as noted in the Figure, shares of FDI by the different trading partners are viewed against the total of FDI minus that attributed to ownership in Hong Kong.
10. Statistics only for Mainland China, from United Nations Conference on Trade and Development (UNCTAD) statistics, available at: http://unctadstat.unctad.org/wds/TableViewer/tableView.aspx?ReportId=96740 [Accessed 1 Apr. 2016].
11. Kolstad and Wiig (2009, p. 4) conclude "accessing large markets and natural resources are important aspects of Chinese outward FDI." The authors caution that fine scrutiny of Chinese FDI data is complicated by difficulties of teasing out flows that move to or originate from Hong Kong for tax and other reasons (p. 4). On the headline-grabbing purchases of well-known properties in the OECD see The Economist (2013).
12. Thayer (2012) details the role that ASEAN has played in the South China Sea disputes from the 1995 Chinese occupation of Filipino-claimed Mischief Reef. See also Sanchita Basu Das (2012) for a snapshot of the issues as they played out in the ASEAN Summit of 2012.
13. See the ASEAN, AEC web page. On the challenges of establishing the AEC, see ADBI (2014) and Das et al. (2013).
14. Schwarz and Villinger (2004, p. 40) discuss the broad problems of different standards and also the lack of scale economies as a result of limited integration and fragmented markets in Southeast Asia.
15. Compare US President Barak Obama: "My top priority as President is making sure more hardworking Americans have a chance to get ahead. That's why we have to make sure the USA—and not countries like China—is the one writing this century's rules for the world's economy" Somanader (2015).
16. See, for instance, the discussion on the ASEAN web site of the ASEAN-China partnership at the 16th Annual ASEAN-China Summit, at: http://www.asean.org/wp-content/uploads/images/archive/23rdASEANSummit/chairmans%20statementfor%20the%2016th%20asean-china%20summit%20-

%20final%203.pdf. China belongs to some formal ASEAN institutions such as the ASEAN Regional Forum (http://aseanregionalforum.asean.org/)

17. Roach (2013) argues that the Chinese economy has become very vulnerable to external shocks. Satyajit Das (2014) provides a good sketch of the evolving vulnerability of the Chinese economy resulting from increasing public and private debt.

Bibliography

Asia Regional Integration Center website. available at: http://aric.adb.org/.

Asian Development Bank (ADB). 2013a. Asian development outlook 2013: Asia's energy challenge. http://www.adb.org/publications/asian-development-outlook-2013-asias-energy-challenge. Accessed 2 May 2015.

———. 2013b. Asian development outlook 2013 update. Governance and public service delivery. http://adb.org/sites/default/files/pub/2013/ado2013-update.pdf. Accessed 13 May 2015.

Asian Development Bank Institute (ADBI). 2014. ASEAN 2030: Toward a borderless economic community. Tokyo, Japan. http://adb.org/sites/default/files/pub/2014/asean-2030-toward-borderless-economic-community.pdf. Accessed 24 Aug 2015.

Association of Southeast Asian Nations (ASEAN DOC). 2002. Declaration on the conduct of parties in the South China Sea. http://www.asean.org/?static_post=declaration-on-the-conduct-of-parties-in-the-south-china-sea-2. Accessed 23 Mar 2016.

———. (ASEAN, AEC). ASEAN Economic Community. http://www.asean.org/communities/asean-economic-community. Accessed 3 Apr. 2016.

———. (ASEAN, APSC). ASEAN Political - Security Community. http://www.asean.org/communities/asean-political-security-community. Accessed 7 Apr. 2015.

ASEAN UP. 2014. Benefits of the ASEAN Economic Community—AEC. ASEAN UP. http://aseanup.com/benefits-asean-economic-community-aec/. Accessed 7 Apr 2015.

Bhattacharya, Abheek. 2014. Japan car makers Skid in China, 3 Nov 2013. *The Wall Street Journal.* http://online.wsj.com/articles/japan-carmakers-skid-in-china-1415009128. Accessed 6 Nov 2014.

Bradsher, Keith. 2010. Amid tension, China blocks vital exports to Japan. The New York Times. 22 Sept 2010. http://www.nytimes.com/2010/09/23/business/global/23rare.html?pagewanted=all&_r=0. Accessed 7 Dec 2014.

———. 2014. Looking beyond China, some companies shift personnel. The New York Times, 9 Sept 2014 http://www.nytimes.com/2014/09/10/business/international/looking-beyond-china-some-companies-shift-personnel.html?_r=0. Accessed 10 Feb 2015.

British Broadcasting Corporation (BBC). 2012. Asean nations fail to reach agreement on South China Sea, 13 July 2012. http://www.bbc.com/news/world-asia-18825148. Accessed 10 Apr 2015.

Choi, Byung-il, and Changyong Rhee. 2014. Future of factory Asia. Asian Development Bank and Korea Economic Research Institute. http://www.adb.org/sites/default/files/publication/42477/future-factory-asia.pdf. Accessed 31 Mar 2016.

Das, Sanchita Basu. 2012. ASEAN: A united front To tackle The South China Sea issue. East Asia Forum. http://www.eastasiaforum.org/2012/05/13/asean-a-united-front-to-tackle-the-south-china-sea-issue/. Accessed 13 May 2012.

Das, Sanchita Basu, Jayant Menon, Rodolfo C. Severino, and Omkar Lal Shrestha (eds). 2013. The ASEAN economic community: A work in progress, Asian Development Bank and Institute of Southeast Asian Studies. http://www.adb.org/sites/default/files/publication/31147/aec-work-progress.pdf. Accessed 1 Jan 2016.

Das, Satyajit. 2014. China's debt vulnerability. EconoMonitor, 9 Apr 2014. http://www.economonitor.com/blog/2014/04/chinas-debt-vulnerability/. Accessed 16 Dec 2014.

De Castro, Renato Cruz. 2013. China's realpolitik approach in the South China Sea dispute: The case of the 2012 Scarborough Shoal stand-off. For Managing Tensions in the South China Sea conference held by the Center for Strategic & International Studies (CSIS) on 5–6 June 2013. http://csis.org/files/attachments/130606_DeCastro_ConferencePaper.pdf. Accessed 23 Nov 2014.

Devonshire-Ellis, Chris. 2014. China-ASEAN wage comparisons and the 70 percent production capacity benchmark. China Briefing, Dezan Shira & Associates, 3 Jun 2014. http://www.china-briefing.com/news/2014/06/03/china-asean-wage-comparisons-70-production-capacity-benchmark.html. Accessed 1 Aug 2015.

Domínguez, Gabriel. 2014. Japan shifting investments from China to ASEAN. Deutsche Welle, 25 Jul 2014. http://www.dw.de/japan-shifting-investments-from-china-to-asean/a-17809068. Accessed 2 Oct 2014.

The Economist. 2013. China's outward investment: The second wave, 26 Oct 2013. http://www.economist.com/news/business/21588369-what-make-chinese-firms-latest-foreign-purchases-second-wave. Accessed 17 Feb 2015.

———. 2014b. Rigged: Vietnam and the South China Sea. 14 Jun 2014. http://www.economist.com/news/asia/21604203-comradely-relations-go-bad-worse-rigged. Accessed 11 Jan 2015.

Francisco, Rosemarie. 2014. Geopolitical tensions could hurt Asia growth, investments: IMF deputy MD. Reuters 21 May 2014. http://www.reuters.com/article/2014/05/21/us-imf-asia-risks-idUSBREA4K0B820140521. Accessed 6 May 2015.

Frost, Ellen L. 2008. *Asia's new regionalism*. Boulder: Lynne Rienner.

Green, David Jay. 2008b. The role of ASEAN economic community as a commitment to policy certainty. *ASEAN Economic Bulletin* August(25/#2): 209–227.

———. 2010. Southeast Asia's policy response to the global economic crisis. *ASEAN Economic Bulletin* April(27/#1): 5–26.

Hayashi, Yuka, and Mayumi Negishi. 2013. Japan's companies Shun China for Southeast Asia. *The Wall Street Journal*, 12 Sept 2013. http://www.wsj.com/articles/SB10001424127887324549004579070820138376020. Accessed 10 Feb 2015.

Hong, Zhao. 2013. China's FDI into Southeast Asia. ISEAS perspective, (Institute of Southeast Asian Studies, Singapore). #08, 31 Jan 2013.

International Monetary Fund. 2014a. Republic of Korea: 2013 Article IV Consultation—Staff Report, Press release and statement by the Executive Director for the Republic of Korea. IMF country report no. 14/101. April 2014. https://www.imf.org/external/pubs/ft/scr/2014/cr14101.pdf. Accessed 6 May 2015.

———. 2014b. Brunei Darussalam: Statistical appendix. IMF country report no. 14/191. July 2014. https://www.imf.org/external/pubs/ft/scr/2014/cr14191.pdf. Accessed 6 May 2015.

———. 2014c. People's Republic of China: 2014 Article IV Consultation—Staff Report; press release; and statement by the Executive Director for the People's Republic of China: IMF country report no. 14/235. July 2014. https://www.imf.org/external/pubs/ft/scr/2014/cr14235.pdf. Accessed 6 May 2015.

———. 2014d. Japan: 2014 Article IV Consultation—Staff Report; and press release. IMF country report no. 14/236. July 2014. https://www.imf.org/external/pubs/ft/scr/2014/cr14236.pdf. Accessed 6 May 2015.

———. 2014e. Philippines: 2014 Article IV Consultation—Staff Report. IMF country report; and press release. No. 14/245. August 2014. https://www.imf.org/external/pubs/ft/scr/2014/cr14245.pdf. Accessed 6 May 2015.

———. 2014f. Vietnam: 2014 Article IV Consultation—Staff Report. IMF country report—Staff report; press release; and statement by the Executive Director for Vietnam. No. 14/311. October 2014. https://www.imf.org/external/pubs/ft/scr/2014/cr14311.pdf. Accessed 6 May 2015.

———. 2015a. Malaysia 2014 Article IV Consultation—Staff report; press release; and statement by the Executive Director for Malaysia. IMF country report no. 15/58. March 2015. https://www.imf.org/external/pubs/ft/scr/2015/cr1558.pdf. Accessed 6 May 2015.

———. (2015b). Indonesia: 2014 Article IV Consultation—Staff Report; Press Release; and Statement by the Executive Director for Indonesia. IMF Country Report No. 15/74. March 2015. https://www.imf.org/external/pubs/ft/scr/2015/cr1574.pdf. Accessed 6 May 2015

———. 2015c. IMF Executive Board Concludes 2015 Article IV Consultation with Brunei Darussalam. Press release no. 15/259. June 4, 2015. https://www.imf.org/external/np/sec/pr/2015/pr15259.htm. Accessed 4 Apr 2016.

Kolstad, Ivar, and Arne Wiig. 2009. What determines Chinese outward FDI? Chr. Michelsen Institute. CMI working paper WP 2009: 3. Bergen, Norway. http://www.cmi.no/publications/publication/?3332=what-determines-chinese-outward-fdi. Accessed 16 Feb 2015.

Roach, Stephen. 2013. China's last soft landing? Project Syndicate, 29 Jan 2013. http://www.project-syndicate.org/commentary/chinese-economic-resilience-is-weakening-by-stephen-s--roach. Accessed 16 Dec 2014.

Roberts, Dexter. 2012. Anti-Japanese protests flare in China over disputed islands. Bloomberg Businessweek, 17 Sept 2012. http://www.businessweek.com/articles/2012-09-17/anti-japanese-protests-flare-in-china-over-disputed-islands. Accessed 7 Dec 2014.

Schwarz, Adam, and Roland Villinger. 2004. Integrating Southeast Asia's economies. *McKinsey Quarterly* (# 1/2004): 36–46. http://unpan1.un.org/intradoc/groups/public/documents/apcity/unpan049156.pdf. Accessed 11 Apr 2015.

Somanader, Tanya. 2015. President Obama: "Writing the rules for 21st century trade". The White House. https://www.whitehouse.gov/blog/2015/02/18/president-obama-writing-rules-21st-century-trade. Accessed 20 Apr 2015.

Supplychains Magazine (Editorial Team). 2012. Diaoyu Island dispute: Japanese supply chain hit hard, 10 Oct 2012. Supplychains Magazine. http://supplychains.com/diaoyu-island-dispute-japanese-supply-chain-hit-hard/. Accessed 8 Dec 2014.

Sussangkarn, Chalongphob. 2012. Toward a functional Chiang Mai initiative. East Asia Forum, 15 May 2012. http://www.eastasiaforum.org/2012/05/15/toward-a-functional-chiang-mai-initiative/. Accessed 15 May 2015.

Thayer, Carlyle A. 2012. ASEAN'S Code of Conduct in the South China Sea: A litmus test for community-building? *The Asia-Pacific Journal* 10(34), no. 4, 20 Aug 2012. http://www.japanfocus.org/-Carlyle_A_-Thayer/3813. Accessed 8 Apr 2015.

Thorbecke, Willem, and Nimesh Salike. 2011. Understanding foreign direct investment in East Asia. ADBI Working Paper 290, June 2011. Tokyo: Asian Development Bank Institute. http://www.adbi.org/working-paper/2011/06/23/4616.foreign.direct.investment.east.asia/. Accessed 1 Apr 2016.

United States Energy Information Administration (US EIA). 2013. South China Sea. http://www.eia.gov/beta/international/regions-topics.cfm?RegionTopicID=SCS. Accessed 23 Jul 2015.

Vu Trong Khanh, and Jenny W. Hsu. 2014. Anger over oil rig turns violent In Vietnam. *The Wall Street Journal*, 14 May 2014. http://www.wsj.com/articles/SB10001424052702303851804579560923647957180. Accessed 17 Mar 2016.

World Bank. 2010. Foreign direct investment – The China story. The World Bank, 16 Jul 2010. http://www.worldbank.org/en/news/feature/2010/07/16/foreign-direct-investment-china-story. Accessed 18 Apr 2015.

World Bank, World Development Indicators. http://databank.worldbank.org/data/reports.aspx?source=world-development-indicators#. Accessed 5 Jul. 2016.

World Trade Organization (WTO)-Institute of Developing Economies (IDE-JETRO). 2011. Trade patterns and global value chains in East Asia: From trade in goods to trade in tasks. https://www.wto.org/english/res_e/booksp_e/stat_tradepat_globvalchains_e.pdf. Accessed 8 Feb 2015.

Zhong, Xinhui. 2013. The gaming among China, the Philippines and the US in the South China Sea disputes. Master Thesis, Development and International Relations, Aalborg University, Denmark. June 2013. http://projekter.aau.dk/projekter/files/76994735/Thesis_final_edition.pdf. Accessed 16 Nov 2014.

5

Hypothetical Rewards, Resources in the South China Sea

Abstract To complete the argument advanced, this chapter examines if the benefits being sought by the South China Sea disputants have been correctly understood. Although the data are less than reliable, a reasonable conclusion is that the rewards that might accrue from *winning* a tussle over the maritime resources are largely illusionary. Gas and oil reserves are likely not present in large amounts in the disputed regions of the South China Sea. Fishery resources are declining and could only be sustainably exploited with the cooperation of the different communities in the region.

Keywords Southeast Asia • Oil and natural gas • Energy reserves • South China Sea • Fishery stocks

This book began with the suggestion that the countries involved in disputes over the South China Sea behave as if the cost to benefit ratio associated with aggressive behavior is low. It then advanced various arguments to the effect that the costs to continued confrontations are large and growing. To complete the argument, we have to examine if the benefits have been correctly or appropriately understood. The review of this issue, centered on the energy and fishery resources, can only be considered tentative because the available data are neither reliable nor precise. Indeed, one is reminded of Mark Twain's warning: "Figures often beguile me, particularly when I have to do the arranging of them myself; in which case

D.J. Green, *The Third Option for the South China Sea*,
DOI 10.1007/978-3-319-40274-1_5

the remark attributed to Disraeli would often apply with justice and force: 'There are three kinds of lies: lies, damned lies and statistics.'"[1] However, given what data there are, the sections below conclude that the rewards that might accrue from *winning* a South China Sea tussle, stemming from the control of these resources, are largely illusionary.

5.1 Energy Resources: There Is No Free Lunch

Energy resources in the South China Sea, oil and natural gas, are sometimes said to be large. Some alternative estimates are given below in Table 5.1. There are other published estimates.[2] The ones listed in Table 5.1 are representative of the literature and are those that can be easily compared to others. Of these, Chinese authorities generally suggest sizeable reserves: Zhong (2013, p. 10) refers to 2012 Chinese Government reports on the South China Sea as bearing oil and gas reserves comparable to the Persian Gulf. The U.S. Energy Information Administration (US EIA 2013), however, is more circumspect, suggesting that the South China Sea contains "approximately 11 billion barrels (bbl) of oil reserves and 190 trillion cubic feet (Tcf) of natural gas reserves". This would put the oil reserves on the order of those of Brazil and natural gas comparable to Venezuela.[3] However, the EIA suggests that most of the presumed reserves are in areas of the South China Sea that are close to the various coastlines and not in the disputed areas. In the more hotly contested areas, the EIA does suggest that there may be significant "undiscovered hydrocarbons" in the

Table 5.1 Alternative estimates of energy resources in the South China Sea

Oil (billion barrels)	*Qualifiers*	*Natural Gas (trillion cubic feet)*	*Qualifiers*	*Source*
130		900		Hong (2012)
125		500		CNOOC (2012) as referenced in Tweed (2015)
7		900		World Bank as referenced in Xu (2014)
11	Proven & probable	190	Proven & probable	EIA (2013) using 2010 data
5–22	Undiscovered	70–290	Undiscovered	EIA (2013) using 2010 data

Notes: *CNOOC* China National Offshore Oil Corporation

Spratly Islands. For the Paracel Islands, the EIA concludes, "Geologic evidence suggests the area does not have significant potential in terms of conventional hydrocarbons."[4]

The wide differences between Chinese sources and those of the EIA are due to many reasons. Rogers (2012, p. 95, fn. 9) notes that the technical conventions of reporting are different: the US estimates focus on likely recoverable resources, not the total existing, while Chinese estimates do not make this adjustment. Rogers further suggests that the (Western) energy industry would typically apply a correction of 90 % and that only 10 % of total reserves would actually be recoverable. This correction would bring the various estimates much closer together.

There are also widely divergent estimates of energy resources in the case of the East China Sea, in the region disputed by China and Japan. O'Shea (2013, pp. 2–3) notes that a published report in 1969 suggested large oil reserves. However, he concludes that few of these can actually be exploited. Natural gas reserves are likely to be present, but "while not insignificant, are marginal in the broader scheme of the energy needs of both states."[5]

In both cases, in the East China Sea and the South China Sea, there are not likely to be huge recoverable energy resources, certainly not of the size to be transformational—to meaningfully change the development path of these countries, especially China. The dramatic fall in crude oil prices in recent times reinforces this conclusion—lower oil prices will likely render uneconomical relatively high-cost deepwater oil and gas production.[6] This accentuates the earlier conclusion that countries may have been underestimating the cost-benefit ratio supporting aggressive behavior in defense of overlapping territorial claims in the west Pacific Ocean.

5.2 Fishery Riches: A Depleting Resource

In addition to oil and gas, the South China Sea is coveted for its fishery resources.[7] All of the littoral countries, as well as some from outside of the region, have an interest in these fish stocks. Rogers (2012, p. 90) notes that "The South China Sea is one of the most biologically diverse marine areas in the world, and some estimates indicate that it is home to nearly 10 percent of the fisheries used for human consumption." Throughout their life cycles, some of the fish stocks *straddle* different regions in this Sea: breeding, birthing, and living in very different areas; sometimes migrating from coastal regions to distant, and in some cases, disputed waters.

The fish stocks are being depleted in part due to unsustainable fishing practices and to coastal development, which has damaged fish breeding grounds. "Since 1970, ever-rising catch volumes have depleted the regional fish stock by 40 % and eradicated over 80 % of large predatory fish."[8] China is the largest producer of fish and related products and has the largest fishing fleet in the region: from 1995 to 2012, the Chinese fishing fleet (motorized) grew more than 60 % (FAO 2012, pp. xvi and 16). Failure to provide for sustainable, inclusive development in the coastal regions will, over time, continue to put pressure on the South China Sea fish stocks, as pollution and the destruction of mangrove forests reduce breeding stocks and at the same time more people seek their livelihood in fishing.

Declining fish stocks encourage fishing boats to move further away from coastal regions. This increases the risk of encounters between fishing boats and maritime security vessels from rival countries in the disputed areas. Indeed, while it is clear that maritime security vessels are agents of government policy, fishing boats also can represent the disputant governments, asserting resource exploitation rights across the breadth of the seas.[9] A good number of the reported incidents in Fig. 2.1, in which countries demonstrate aggressive behavior in defense of their rights in the maritime areas, involve confrontations between naval or security vessels and fishing boats.

Some of these clashes between security vessels and fishing boats are in the name of protecting or husbanding the fishery resources, of enforcing laws regarding sustainable fishing practices. Taking this effort at face value—that nations want to enforce sustainable fishing policy regimes—there is likely no way that individual nations can succeed in uncoordinated efforts. The geography of a huge shallow sea, surrounded by the world's largest archipelagoes dotted with many poor villages dependent upon fishing, presents a nearly insurmountable policing problem. Policing and sustainable resource management will require cooperative solutions, would need the collaboration of the different communities surrounding the South China Sea.

In summary, the fishing resources of the South China Sea are not negligible; however, they are not likely to be a treasure trove that can be captured and surely exploited by any particular territorial claimant. Sustainable resource exploitation will require collective and coordinated coastal and marine efforts. As discussed more fully below, regional cooperation provides a potential avenue for successful management of fishing stocks, although one strewn with challenges.

Notes

1. From Mark Twain's Own Autobiography (Twain 1904).
2. For instance, Khemakorn (2006, p. 16) refers to "a 1995 study by Russia's Research Institute of Geology of Foreign Countries [that] estimates that an equivalent of 6 billion barrels of oil might be located in the Spratly Islands area, of which 70 % would be natural gas."
3. The US EIA estimated that in 2015 Brazil had crude oil reserves of approximately 15 billion barrels, while Venezuela has 197 trillion cubic feet of natural gas. (US EIA web site http://www.eia.gov/beta/, accessed April 4, 2016).
4. US EIA (2013). The picture drawn by the EIA is consistent with recent Chinese exploration that reports finding significant natural gas reserves close to Hainan Island, territory unclaimed by others, but finding less in the disputed Paracel Islands. (Tiezzi, 2014)
5. O'Shea (2013, p. 3). O'Shea also pointedly concludes concerning the conflicting claims in the East China Sea between China and Japan: "Indeed, the disruption to bilateral trade and resulting economic losses caused by the dispute outweigh the value of the deposits themselves."
6. The price for the benchmark West Texas Intermediate (WTI) crude oil price in 2015 was almost 48 % below the average for 2014. (US EIA, available https://www.eia.gov/dnav/pet/PET_PRI_SPT_S1_A.htm)
7. "The right to harvest the South China Sea's resources is one of the main drivers behind territorial disputes." (Pejsova 2014, p. 1) Khemakorn (2006) and Rogers (2012) provide useful discussions of the issues surrounding the fishery resources in the South China Sea.
8. Pejsova (2014, p. 1). See also Baker (2016) and Khemakorn (2006, p. 32), who notes "2/3 of the major fish species are overexploited."
9. See Fravel (2012, especially pp. 37–38) and Rogers (2012, p. 89).

Bibliography

Baker, Rodger. 2016. Fish: The overlooked destabilizer in the South China Sea, 12 Feb. 2106. Stratfor. https://www.stratfor.com/analysis/fish-overlooked-destabilizer-south-china-sea. Accessed 21 Mar 2016.

Food and Agriculture Organization (FAO). 2012. FAO yearbook. Fishery and aquaculture statistics. FAO Fisheries and Aquaculture Department. http://www.fao.org/3/a-i3740t/index.html. Accessed 5 Apr 2016.

Fravel, M. Taylor. 2012. Maritime security in the South China Sea and the competition over maritime rights. In *Cooperation from strength the United States, China and the South China Sea*, ed. P. Cronin. Washington, DC: Center for a New American Security, 31–50. http://www.cnas.org/sites/default/files/publications-pdf/CNAS. Accessed 16 May 2015.

Hong, Zhao. 2012. Sino-Philippines relations: Moving beyond South China Sea dispute? *The Journal of East Asian Affairs* 26(2(Fall/Winter)): 57–76.

Khemakorn, Pakjuta. 2006. Sustainable management of pelagic fisheries in the South China Sea Region, Nov 2006. http://www.un.org/depts/los/nippon/unnff_programme_home/fellows_pages/fellows_papers/khemakorn_0607_thailand.pdf. Accessed 17 May 2015.

O'Shea, Paul. 2013. Territorial disputes in Northeast Asia: A primer. Italian Institute for International Political Studies Analysis (ISPI) No. 182, June 2013. http://www.ispionline.it/sites/default/files/pubblicazioni/analysis_182_2013.pdf. Accessed 8 Dec 2014.

Pejsova, Eva. 2014. The South China Sea's commons: Behind and beyond sovereignty disputes. European Union Institute for Security Studies, June 2014. http://www.iss.europa.eu/uploads/media/Alert_30_South_China_Sea.pdf. Accessed 17 May 2015.

Rogers, Will. 2012. The role of natural resources in the South China Sea. In *Cooperation from strength the United States, China and the South China Sea*, ed. P. Cronin, 83–97. Washington, DC: Center for a New American Security. http://www.cnas.org/sites/default/files/publications-pdf/CNAS. Accessed 16 May 2015.

Tiezzi, Shannon. 2014. China discovers gas field in the South China Sea. The Diplomat, 16 Sept 2014. http://thediplomat.com/2014/09/china-discovers-gas. Accessed 16 May 2015.

Twain. 1904. From *Mark Twain's own autobiography*. Directory of Mark Twain's maxims, quotations, and various opinions. http://www.twainquotes.com/Lies.html. Accessed 4 Apr 2016.

Tweed, David. 2015. What do weak oil prices mean for the South China Sea? Bloomberg.com, 20 Jan 2015. http://www.bloomberg.com/news/articles/2015-01-20/all-about-the-base-oil. Accessed 17 May 2015.

United States Energy Information Administration (US EIA). 2013. South China Sea. http://www.eia.gov/beta/international/regions-topics.cfm?RegionTopicID=SCS. Accessed 23 Jul 2015.

Xu, Beina. 2014. South China Sea tensions. Council on Foreign Relations—CFR Backgrounders, 14 May 2014. http://www.cfr.org/china/south-china-sea-tensions/p29790. Accessed 22 May 2015.

Zhong, Xinhui. 2013. The gaming among China, the Philippines and the US in the South China Sea disputes. Master Thesis, Development and International Relations, Aalborg University, Denmark. June 2013. http://projekter.aau.dk/projekter/files/76994735/Thesis_final_edition.pdf. Accessed 16 Nov 2014.

6

Broader Issues in the West Pacific

Abstract This chapter provides a perspective on the earlier discussion on economic interests by reviewing a broader set of concerns that motivate national behavior in the South China Sea. These range from national security to the need to reduce the spread of infectious diseases, control piracy, and mitigate the impact of natural disasters. The chapter foreshadows later discussions on collaborative initiatives, as many of these concerns can be met only through regional cooperation. This section also reviews the interests and impact of non-regional players, particularly the USA and Australia. These other players, responding to their own needs and concerns, heighten the risk of confrontation sparking conflict: the internationalization of the South China Sea raises risks to all.

Keywords Southeast Asia • National security • United States • Australia • South China Sea • Conflict

6.1 National Security and Beyond

For many observers, the perspective of this book will seem narrow as it has largely ignored security issues. Disputes over the maritime rights in the South China Sea do need to be seen in the light of broader national interests, many related to security. If behavior in the various disputes is linked to an assessment of the costs and benefits that might accrue,

D.J. Green, *The Third Option for the South China Sea*,
DOI 10.1007/978-3-319-40274-1_6

we surely need to look beyond narrow economic interests, as important as they may be. This acknowledgment does not ease the analytic problem, as each country faces myriad concerns.

Table 6.1 attempts to provide a listing of some of these varied national issues, many culled from the large literature on the South China Sea. The list deliberately ignores a number of economic issues, such as mitigating cross-border contagion from trade shocks. These subjects were dealt within the sections above. By no means is this an exhaustive list. Anyone can come up with other issues that should or could make this list, like slowing or mitigating the impact of climate change or establishing peaceful dispute resolution mechanisms. The operative screen was to choose issues that are acknowledged in the literature. (Appendix B provides a link between this listing and some of the studies on the subject.)

While these concerns are noted in a simple list, many are interrelated. Progress on smuggling might aid in reducing piracy and lessening the opportunities to fish illegally or in unsustainable fashions. Similarly, reducing water pollution will likely aid in protecting fish stocks. Even security is not a standalone issue: Frost, for example, discusses the interrelationship in East Asia between security and economic concerns (2008, p. 199).

Table 6.1 also attempts to provide the structure for these concerns from the standpoint of how they may be resolved or progress made, particularly as to whether, under current political conditions, cooperative behavior

Table 6.1 Policy issues in the South China Sea disputes

Issues:	**Issues**
Involving zero-sum relationships	Providing strategic space for security forces Enforcing sovereignty, control of territory Projecting an image of strength
Largely within the control of a single state	Accessing mineral, especially energy, resources Disaster relief Protecting the environment, reducing water pollution Counterterrorism
Requiring or enhanced by collaborative actions	Controlling illegal trade, trafficking in people, and smuggling of controlled goods and substances; enforcing sanitary and phytosanitary customs rules Combating piracy Reducing illegal, unsustainable fishing Ensuring freedom of transit and travel Preventing the spread of infectious diseases

Note: See Appendix 2 for sources

would be useful. For example, one issue listed, enforcing sanitary and phytosanitary rules, is noted as one that requires or would be enhanced by collaborative actions. Formally most of the work in this area is within a country's border posts where the custom and immigration work is done to enforce these rules. However, common regional rules, transparent procedures, and cooperation with trading partners can enhance any one country's efforts.

A closely related issue, preventing the spread of infectious diseases, is also something that takes cooperative efforts. A collapse of public health programs in one country can lead to a spread of disease to a regional neighbor in spite of that individual country's efforts to seal borders to infected produce, people, or animals. The clear experience, learned at some cost but well understood by most Asian governments, is that when a disease such as severe acute respiratory syndrome (SARS) or Avian Influenza appears in one country, neighboring countries or trading partners are at risk. While the bulk of a nation's response to the threat of disease must be directed within its borders, the costs fall and the efficacy of protective efforts rise with coordinated efforts to limit the spreading of disease across borders.

Similarly, husbanding fishing stocks requires both national and internationally coordinated programs. Fish, as is often noted, do not carry passports and may spawn and spend the early part of their lives in the coastal area of one nation, traveling to seas near another later in life. Uncoordinated rules and the enforcement thereof concerning the taking of fish in one country may affect the industry in another. Conversely, coordinated efforts might reward both with the combined benefits exceeding what each individual country could expect from uncoordinated policy regimes. In this case, what first appears to be a zero-sum game with a fixed amount of resources to be divided between two contestants can become a positive sum game as collaboration improves the opportunities for all.

Not all issues encourage cooperative behavior, at least given the present and likely regional political environment. For instance, every country values national security, which is often perceived as providing for strategic space for military forces. Occupying even rocky, semi-submerged islets in the South China Sea, denying these spaces to others, might be seen as a gain for national security. In this case, we are likely dealing with a zero-sum situation; either you or I stand on the rocky outcropping, not both of us. There are collaborative solutions for meeting national security needs, but they are qualitatively different than providing for regional natural disaster management.

In some instances, the issues in Table 6.1 cannot be neatly distinguished from one another: enforcing sovereignty, control of territory, for example, is one way to project an image of strength. Moreover, some issues are what might be termed secondary or derived national concerns. Presumably, no one cares about mineral resources in and of themselves; rather, the interest is from an underlying goal of ensuring for economic growth. This is relevant in that there may be the possibility for defusing conflict, reducing concerns over some particular issue in return for obtaining progress toward others or toward ensuring economic growth generally.

The distinction between competitive and collaborative or cooperative issues suggested in Table 6.1 is clearly not one that is fixed in stone: changing economic institutions, changing political currents, experiential learning, all can alter the nature of how policymakers view the potential for collaborative solution to issues.[1] The distinctions made in this section are made less from a desire to speak ex cathedra in detail about any particular issue, than to motivate the discussion later in the book as to the nature of the collaborative institutions that would be required to ease tensions in the South China Sea.

6.2 Internationalization, the Involvement of Non-Regional Players

The last section attempted to provide a wider perspective on the earlier discussion, which somewhat narrowly framed an analysis of the tensions in the South China Sea from the standpoint of economic concerns. Here we similarly expand the focus, looking at the risks of internationalization of what have largely been regional problems. In this case, the risks are those of having non-regional players such as the USA become heavily involved. The possible role of other countries, such as Australia will also be noted, but the USA is the one potential entrant that has the ability for real game changing, for better or worse.

President Barak Obama used a meeting with ASEAN leaders in early 2016 to express US concerns. He said: "We discussed the need for tangible steps in the South China Sea to lower tensions, including a halt to further reclamation, new construction and militarization of disputed areas. Freedom of navigation must be upheld and lawful commerce should not be impeded"(US Press Secretary 2016). He further stated the country's intentions to meet these concerns through specific actions: "I reiterated that the United States will continue to fly, sail, and operate wherever international law allows, and we will support the right of all countries to do the same. We will continue to help our allies and partners strengthen their maritime capabilities."

This policy commitment confirmed many earlier statements by government officials. David Shear, Assistant Secretary of Defense for Asian & Pacific Security Affairs, speaking before the U.S. Senate Committee on Foreign Relations in mid-2015, stated the country's aims: "peaceful resolution of disputes, freedom of navigation and overflight and other internationally lawful uses of the sea related to these freedoms, unimpeded lawful commerce, respect for international law, and the maintenance of peace and stability" (2015, p. 4). Within these relatively high-level goals, he notes that, "First, we are committed to deterring coercion and aggression and thereby reinforcing the stability of the Asia-Pacific region" (2015, p. 5). In theory, as all nations have committed what, to other claimants, are aggressive acts, this would put the USA squarely in the middle of a number of bilateral disputes in the West Pacific Ocean. In fact, Shear notes immediately that with Japan, the Philippines, and South Korea, the USA has been able to "refresh and modernize our long-standing alliances" (p. 5). Of the active disputants, this would leave China and Viet Nam bereft of US support.

Indeed, although Shear's testimony notes that Viet Nam was, until recently, the "most active claimant" (2015, p. 3), it is clear that the real concerns today are the actions of China. Succinctly, "We are concerned that the scope and nature of China's actions have the potential to disrupt regional security" (p. 3). This is echoed in the CSIS (2016, p. 19) study, which, in describing the strategic dilemma of the USA argues that by 2030, with the growing economic and military power of China, "the South China Sea will be virtually a Chinese lake." Overwhelmingly the concern of the USA appears to focus on China.[2]

What this preoccupation might mean for the region, how the USA might become involved in the South China Sea disputes, is still to unfold. There is the strong suggestion, however, that, the USA would take sides. President Obama's statement noted above is only one of many that would have the USA help "allies and partners."[3] As this manuscript is being finalized, reports that the USA has placed forces in the Philippines to support joint military operations in the South China Sea is a strong indication that the USA might be seen as a Filipino ally (Whaley 2016).

From the standpoint of our earlier discussion, from seeing the confrontations in the South China Sea as a series of games of chicken, the possible aligning of the USA with one or another country does not fundamentally change our analysis. Within the simple game, we could introduce the possible engagement of the USA into, say, the dispute between China and the Philippines, by adjusting the cells of the game framework. The USA as an ally certainly raises the possible costs of aggression: US military forces are undoubtedly formidable, and conflict or even the threat of conflict could involve massive economic repercussions.

The involvement of the USA as an ally to one of the disputants would internationalize the costs and impacts of real conflict. However, while there is certainly the strong suggestion that the USA might enter militarily in the event of a conflict, especially in the South China Sea there has been no concrete commitment to honor any particular country's claims, even with respect to the Philippines (ICG 2012, p. 25). Without this, it is difficult to see the possibility of the USA acting as an ally to have definite impact on the nature of current play.

What does have the potential for changing the nature of the conflict in the South China Sea is recognizing that the USA has entered the field as a player in its own right. President Obama made this clear by saying and confirming that US military planes and ships will continue to fly and sail in the region, undeterred by claims of ownership. Over a decade ago, the undertaking of the USA in the South China Sea of "[a]erial reconnaissance and surveillance activities" brought military forces of China and the USA literally into collision (CSIS 2016, pp. 42–43).

The policy of the USA, centering on the right of free transit, is not the same as, for instance, the rights of Filipino vessels to take fish throughout the South China Sea. Thus, it is probably better to see the USA as an additional claimant of rights to the South China Sea, one that seeks to deny, especially, Chinese domination. As a player in its own right, the USA is initiating a game with China similar to that played by the other South China Sea disputants. From this standpoint, Cronin (2014) discusses explicitly how US policy could raise the costs to the Chinese government of aggressive behavior.[4]

China and the USA now collectively face real risks as each country asserts their rights, defends their interests. At risk especially is the economic partnership that has grown over the past two decades, a partnership that is one of the most important that each country has and one that is vitally important to the global economy. In 2015, China was the

USA's largest supplier of merchandise imports, the third largest export destination.[5] Similarly in 2013, the USA was the largest merchandise export market for China and one of the larger import providers.[6] The economies are equally intertwined on the capital accounts. Considerable public attention, for instance, has been paid to Chinese ownership of US government debt—in January 2016, Chinese and Hong Kong residents held more than $1.4 trillion of U.S. Treasury securities, more than 23.3 % of the total public debt held abroad.[7]

It is hard to believe that confrontations between China and the USA will not generally raise the odds of actual conflict in the South China Sea. It would seem axiomatic that the more players, the more games of chicken there are and the more likely that one of them would trigger trade- and development-destroying conflict.

Besides the USA, several other countries have clear interests in and could become involved in South China Sea disputes. Australia, Japan, and South Korea all have economies that depend critically on sea-borne trade that passes through the South China Sea. Japan and South Korea could well see strong linkages between events in the South China Sea and prospects for them in the northern maritime disputes. Finally, India is a non-regional country that has recently signaled an interest in the South China Sea by taking part in nearby naval exercises with Japan and the USA (Friedman 2016). In each case, while every country has different and indeed ambiguous concerns and interests, the involvement of non-regional players is likely to raise the likelihood and costs of conflict.

Looking more closely at Australia can help elaborate on these points. Much of Australia's merchandise trade moves through the South China Sea. The Canberra Government has reacted to the militarization of the South China Sea, particularly the building of bases by China, through boosting the defense budget and expanding cooperation with the US military (Taylor 2016). Australia's policies, echoing those of the USA, emphasize that the waterways of the South China Sea should be open to international air and sea transit. Indeed, there are reports that Australia mounted flights by military aircraft to assert this right.[8]

Australian policymakers, however, like those of other countries involved in the South China Sea disputes, are grappling with conflicted interests. On one hand, there is the impulse to defend maritime rights, in this case, the unrestricted right of transit; on the other hand, there is the concern to minimize risks to existing economic relations, especially trade. Figure 6.1 shows the increasing importance of China to the Australian

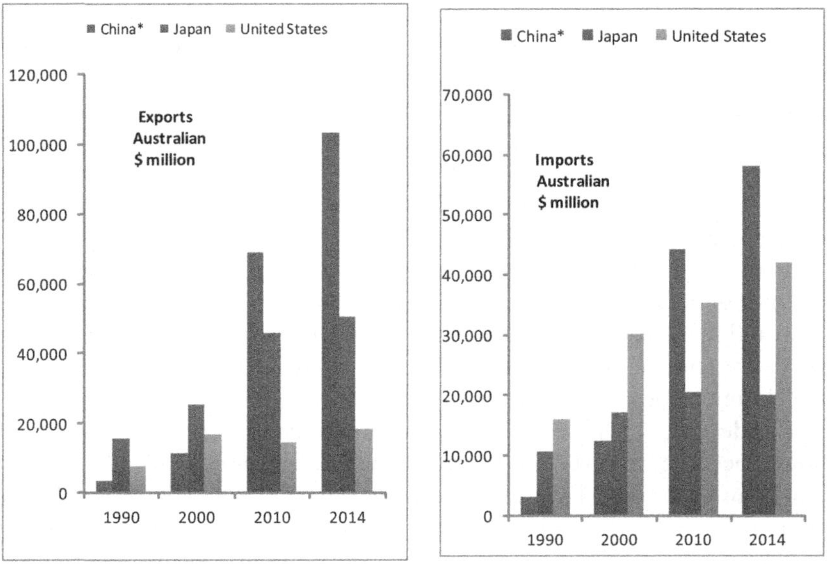

Fig. 6.1 Australian exports and imports.
Notes: China* includes Hong Kong and Macau. Source: Australia, Department of Foreign Affairs and Trade, available at: http://dfat.gov.au/trade/resources/trade-statistics/Pages/trade-time-series-data.aspx

economy. Whereas in 1990, China absorbed 5.3 % of Australia's exports of goods and services, by 2014, this had grown to 31.6 %. The picture from the imports side is not that much different—China is Australia's biggest trading partner.

Australia and the other countries mentioned above all have an interest in preventing any country (meaning China) from controlling transit through the South China Sea; however, much of the trade they worry about, especially for Australia is with China. Thus, like all countries involved in the maritime disputes, they have every reason to make their point hoping not to push the issue to outright conflict. Is this hope reasonable? The underlying disputes occasioning repeated confrontations, the militarization of the islands, and the increasing involvement in the disputes of non-regional countries all point in the direction of increased risk of international conflict. With this established, we turn to the potential for regional cooperation to provide an alternative future.

Notes

1. Hayton (2014, p. 150) suggests that easing of tensions over sovereignty might enhance individual and collective efforts toward energy security.
2. Statements testifying to this conclusion can be found across the literature on the South China Sea. Cronin and Kaplan (2012, p. 5) write "American interests are increasingly at risk in the South China Sea due to the economic and military rise of China and concerns about its willingness to uphold existing legal norms." Erickson and Strange (2014) conclude that China "is the only South China Sea claimant that is potentially capable of establishing de facto air and sea denial over tiny islet networks in a maritime setting as vast as the Spratly archipelago." The CSIS (2016, p.10) study similarly notes: "The course charted by China's reemergence as a great power over the next few decades represents the primary strategic challenge for the United States."
3. "If confrontation were to involve Japan in the East China Sea or the Philippines in the South China Sea, the United States would be obligated to consider military action under defense treaties" (CFR 2016). A key word in this statement is *consider*, see the text below.
4. "This report is the first in a series designed to address strategies for imposing costs on bad behavior in maritime Asia" Cronin (2014, p. 4). A note of warning is also given: "Clearly it is not wise to deny an accretion of Chinese influence over its near seas if it comes at the price of war" (Cronin 2014, p. 15).
5. United States Census Bureau (2015), Top Trading Partners—December 2015.
6. National Bureau of Statistics of China (2014), Table 11-6 Value of Imports and Exports by Country (Region) of Origin/Destination.
7. U.S. Treasury (2016), Recent data for all countries, Table 3D: U.S. Securities Held by Foreign Residents.
8. BBC (2015b). Stratfor (2016) suggests Japan is also edging toward a military "presence" in the South China Sea.

Bibliography

British Broadcasting Corporation (BBC). 2015b. Australia conducting 'freedom of navigation' flights In South China Sea—BBC News, 15 Dec 2015. http://www.bbc.com/news/world-australia-35099445. Accessed 19 Mar 2016.

Center for Strategic and International Studies (CSIS). 2016. Asia-Pacific rebalance 2025 capabilities, presence, and partnerships: An independent review of U.S. defense strategy in the Asia-Pacific. http://csis.org/files/publication/160119_Green_AsiaPacificRebalance2025_Web_0.pdf. Accessed 11 Mar 2016.

(China) National Bureau of Statistics of China. 2014. China statistical yearbook-2014. http://www.stats.gov.cn/tjsj/ndsj/2014/indexeh.htm. Accessed 20 Mar 2016.

Council on Foreign Relations (CFR, 2016). China's Maritime Disputes. A CFR InfoGuide Presentation. http://www.cfr.org/asia-and-pacific/chinas-maritime-disputes/p31345#!/p31345. Accessed March 6, 2016.

Cronin, Patrick M. 2014. The challenge of responding to maritime coercion. Washington, DC: Center for a New American Security, Sept 2014. http://www.cnas.org/Challenge-Responding-to-Maritime-Coercion#.Vu8M0xIrJDU. Accessed 20 Mar 2016.

Cronin, Patrick M., and Robert D. Kaplan. 2012. Cooperation from strength: U.S. strategy and the South China Sea. In *Cooperation from strength: The United States, China and the South China Sea*, ed. Cronin, 3–30. Washington, DC: Center for a New American Security. http://www.cnas.org/sites/default/files/publications-pdf/CNAS. Accessed 16 May 2015.

Erickson, Andrew S., and Austin Strange. 2014. Pandora's Sandbox: China's Island-building strategy in the South China Sea. *Foreign Affairs* 13 Jul 2014. https://www.foreignaffairs.com/articles/china/2014-07-13/pandoras-sandbox. Accessed 6 Mar 2016.

Friedman, George. 2016. The significance of US, Indian and Japanese naval exercises. Geopolitical Futures, 4 Mar 2016. https://geopoliticalfutures.com/the-significance-of-us-indian-and-japanese-naval-exercises/. Accessed 20 Mar 2016.

Frost, Ellen L. 2008. *Asia's new regionalism*. Boulder: Lynne Rienner.

Hayton, Bill. 2014. *The South China Sea: The struggle for power in Asia*. New Haven: Yale University Press.

International Crisis Group (ICG). 2012. Stirring up the South China Sea (II): Regional responses, report N 229, 24 July 2012. http://www.crisisgroup.org/~/media/Files/asia/north-east-asia/229-stirring-up-the-south-china-sea-ii-regional-responses. Accessed 22 May 2015.

Shear, David. 2015. Statement of David Shear Assistant Secretary of Defense for Asian & Pacific Security Affairs before the Senate Committee on Foreign Relations, May 13, 2015. U.S. Senate Committee on Foreign Relations. http://www.foreign.senate.gov/imo/media/doc/051315_Shear_Testimony.pdf. Accessed 6 Mar 2016.

Stratfor Global Intelligence (Stratfor). 2016. Japan wades further into the South China Sea dispute, 12 Jan 2016. Geopolitical Diary. https://www.stratfor.com/geopolitical-diary/japan-wades-further-south-china-sea-dispute. Accessed 21 Mar 2016.

Taylor, Rob. 2016. Australia takes steps to counter China's rising military power. *The Wall Street Journal*, 24 Feb 2016. http://www.wsj.com/articles/australia-takes-steps-to-counter-chinas-island-building-1456366660. Accessed 19 Mar 2016.

United States Census Bureau (U.S. Census Bureau). 2015. Foreign trade – U.S. top trading partners. *Census.Gov*. https://www.census.gov/foreign-trade/statistics/highlights/top/top1512yr.html. Accessed 20 Mar 2016.

United States Department of Treasury (U.S. Treasury). 2016. Treasury International Capital (TIC) System – Home Page. https://www.treasury.gov/resource-center/data-chart-center/tic/Pages/index.aspx. Accessed 20 Mar 2016.

United States, Office of the Press Secretary, the White House (U.S. Press Secretary). 2016. Remarks by President Obama At U.S.-ASEAN press conference, 16 Feb 2016. Home, Briefing room, speeches & remarks. https://www.whitehouse.gov/the-press-office/2016/02/16/remarks-president-obama-us-asean-press-conference. Accessed 20 Mar 2016.

Whaley, F. 2016. *U.S. and Philippines bolster air and sea patrols in South China Sea*. New York Times, 14 Apr 2016. http://www.nytimes.com/2016/04/15/world/asia/south-china-sea-philippines-us-naval-patrols.html?_r=0. Accessed 19 Apr 2016.

7

Regional Cooperation as the Third Option: A Modified Game

Abstract This chapter introduces into the game theory model used earlier, the possibility of regional cooperation as a third alternative to the present behavior of intermittent passive and aggressive pursuit of rival maritime claims. The discussion reveals that regional cooperation must be more than simply sharing existing resources; the exercise must provide some real additionality to the current contest. Offering more than might be obtained by conflict is likely the only path to the existing, unstable behavior.

Keywords Game theory • Conflict • South China Sea • Regional and international cooperation

Table 6.1 suggests that at least some issues motivating conflict in the South China Sea could be resolved or addressed within a cooperative framework. We examine the nature of regional cooperation needed in this section, first returning to the game theory framework used above. In the earlier model, two possible behaviors were allowed, either passive or aggressive defense of public claims to maritime regions of the South China Sea. Regional cooperation provides a third option. This is illustrated in Table 7.1, adding to our earlier game a row and column, allowing for a collaborative policy stance.

D.J. Green, *The Third Option for the South China Sea*,
DOI 10.1007/978-3-319-40274-1_7

Country and Behavior

B \ A	Collaborative	Passive	Aggressive
Collaborative	Z* / Z*	X / X	2X / 0
Passive	X / X	X / X	2X / 0
Aggressive	0 / 2X	0 / 2X	-Y* / -Y*

Z*, Y* > X > 1

Table 7.1 Costs and benefits in an expanded game

The elements in the new cells reflect the following assumptions:

- The potential benefits from simultaneously offering to collaborate, Z*, are substantial and equal for the two countries.
- If you offer to collaborate, but your counterpart acts otherwise, it is the same as if you had simply responded in a passive fashion.

Collaborative behavior, offering to cooperate, is then at least as beneficial as passively asserting your claims—collaborative behavior *weakly* dominates passive behavior. Simplifying the argument, we can eliminate passive behavior as a choice in favor of always behaving either collaboratively or aggressively. This yields the left-hand side 2 × 2 matrix given below (Table 7.2a).

The right-hand side matrix (Table 7.2b) comes from normalizing on X. For clearly identifiable values of Z and Y there is a Nash equilibrium that is Pareto optimal: collaborative solutions could be the best for both players. This is fairly intuitive: if the gains from joint collaborative behavior (Z) are more than twice the potential gains from separately, passively exploiting the region, if they are larger than the potential gains from aggressive behavior, then the countries will cooperate. Put in the negative, the persistent, if uneven, resort to aggressive behavior by all South China Sea claimants suggests that policymakers do not see that collaborative behavior is potentially rewarding: regional cooperation is not yet a game-changer.

Country and Behavior

B \ A	Collaborative	Aggressive
Collaborative	Z* / Z*	2X / 0
Aggressive	0 / 2X	−Y* / −Y*

Z*, Y* > X > 1

Country and Behavior

B \ A	Collaborative	Aggressive
Collaborative	Z / Z	2 / 0
Aggressive	0 / 2	−Y / −Y

Y=Y*/X > 1; Z=Z*/X > 1

Table 7.2 (**a**) Costs and benefits (**b**) Normalizing on X

"The key would be to find a way to collectively develop and share the resources, without renouncing respective claims—to agree to share now and (maybe) fight later" (Green 2013). One issue to emphasize: a regional program that provides for collaborative realization of current disputant goals would have to be more than simply a resource sharing arrangement. Resource sharing is essentially a zero-sum game, which is how countries currently see the situation and they have so far opted to try and get more for themselves through determined if inconsistent aggressive behavior. A regional program would need to be a positive sum game and convincingly to be so.

This is also why appeals to cooperate in areas such as fishery resource management or the suppression of piracy are likely to fall on stubbornly deaf ears.[1] For instance, the Declaration on the Conduct of Parties in the South China Sea—intended to be a central effort to ease tensions over rival maritime claims—suggests a number of areas for mutual cooperation:

(a) Marine environmental protection
(b) Marine scientific research
(c) Safety of navigation and communication at sea
(d) Search and rescue operation
(e) Combating transnational crime, including but not limited to trafficking in illicit drugs, piracy and armed robbery at sea, and illegal traffic in arms (ASEAN DOC 2002)

But these types of collaborative exercises will not appear to be worth what is thought to be given up—the promised treasures of solitary control of the seas. Later, it will be argued that efforts in these areas can play a useful role, but only in the context of larger, broader and potentially more worthwhile regional cooperation initiatives.

Successful regional cooperation can reduce the attractiveness of armed conflict, but outside of the cartoon-like simplicity of game theory, it does not do so by substituting harmony for dispute. One problem is that one person does not accomplish national decision-making at one time for all-time; national decision-making reflects actions by a range of players acting at different times and places and, critically, with interests that are not common. In the South China Sea, the career interests of a maritime security officer from country A facing a fishing vessel from country B are far removed from those of the manager or workers of the manufacturing plants relying on smooth economic relations between the two nations. At a national level, the interests of the defense ministry will be very different from officials in the ministry of trade or development. Regional cooperation would need to materially strengthen the positions of some actors on the national stage.

Highlighting that regional cooperation is not a simple panacea, paradoxically regional cooperation can create a host of new areas of contention between participating countries. As discussed below, the experience of regional cooperation in Southeast Asia, particularly with those initiatives directed toward quickening trade and hastening economic growth, suggests that there will be many possible areas of contention that need to be resolved. Where are infrastructure investments to be located? Whose tariffs tumble the most? Which country's customs procedures provide the regional model? What capital hosts the Leaders' Summit? How do we publically acknowledge and address capacity weaknesses? Regional programs can widen the scope for interaction between countries, sometimes leading to increased opportunities for tension as well as problem solving. Overall, however, widening the game through regional cooperation should raise interest in maintaining an environment conducive to trade, travel, and transit, reducing the risks of resorting to the use of force.

Note

1. Stoa (2015) argues that "A framework for coral reef protection and fisheries management" might ease tensions over security and energy resource use. Later in the text we argue that we will likely need a larger program of cooperation to find the space for progress in this and other important areas.

Bibliography

Association of Southeast Asian Nations (ASEAN DOC). 2002. Declaration on the conduct of parties in the South China Sea. http://www.asean.org/?static_post=declaration-on-the-conduct-of-parties-in-the-south-china-sea-2. Accessed 23 Mar 2016.

Green, David Jay. 2013. Fighting over the West Philippine Sea is so 17th century. *Business Mirror*, Manila, 30 June 2013.

Stoa, Ryan. 2015. Environmental peacebuilding in the South China Sea. 1 & 9 May 2015 [Blog] http://ryanstoa.com/blog?category=South+China+Sea. Accessed 25 Mar 2016.

8

The Experiences of Existing Regional Cooperation Initiatives

Abstract This chapter reviews the Southeast Asian experience in regional cooperation initiatives. Particularly relevant are the Greater Mekong Subregion Economic Cooperation Program on the mainland of Southeast Asia and the Indonesia-Malaysia-Singapore—Growth Triangle. These initiatives have stimulated growth and likely have encouraged the participating countries to maintain smooth economic relations. Other regional programs have been less successful. Lessons learned include (i) the utility of exploiting cross-border resource complementarity and (ii) the importance of trade-encouraging infrastructure investment. These issues have been particularly difficult to address in programs in the archipelagic countries, where the economies show little complementarity and where physical connectivity has been limited, hindering trade and traffic.

Keywords Greater Mekong Subregion (GMS) • Brunei-Indonesia-Malaysia-Philippines—East ASEAN Growth Area (BIMP-EAGA) • Regional cooperation • Singapore • Trade and transport facilitation

As a working definition, regional economic cooperation allows countries to seek collaborative solutions to development problems including bringing infrastructure to borderlands, harmonizing procedures and technical standards that affect international trade in goods and services, providing

D.J. Green, *The Third Option for the South China Sea*,
DOI 10.1007/978-3-319-40274-1_8

scope for cross-border investment and economies of scale in production, and addressing cross-border environmental and security issues. The question then to be answered is, "Can this type of cooperation meaningfully provide a third option for the South China Sea disputants?" We approach this by reviewing the rich set of experiences of these countries with economic cooperation initiatives and attempt to focus on two issues: (i) can cooperation in the South China Sea successfully and materially raise economic growth, and (ii) would the benefits be large enough to discourage conflict over the maritime territory?[1]

Regional cooperation broadly defined exists in many forms among the South China Sea claimants. Indeed, there is such an array of organizations, each generally known by an acronym, that the term *alphabet soup* can be applied (Green 2008a). The multiplicity of initiatives, from Asia-Pacific Economic Cooperation (APEC) to Zone of Peace, Freedom and Neutrality (ZOPFAN) testifies to an undercurrent of efforts by national governments to seek collective solutions to common problems.[2] Some of these forums are designed to encourage trade and thus growth, for example, APEC. But, worthwhile as these initiatives are, they have thus far not deterred the countries involved from behaving aggressively in pursuit of their South China Sea interests. For example, all ASEAN member states participate, along with China, in the Regional Comprehensive Economic Partnership (RCEP) negotiations. However, it is unlikely that this initiative will present strong opportunities for growth of the magnitude that could alter the present behavior in the South China Sea.

Table 8.1 provides a summary of those regional cooperation initiatives most relevant for the purposes of this book—those providing for broad economic development programs and involving some of the South China Sea disputants.[3]

The focus here is on economic development cooperation because it is only in this general sphere of action that sufficient benefits could be created to alter the perception of the costs and benefits of current behavior in the South China Sea. Cooperating on sharing fishery resources or on combatting smuggling are certainly worthwhile activities, but they do not stack up well against security concerns; raising growth prospects might.

The initiatives noted in Table 8.1 attempt to create real growth opportunities by acknowledging that borders can be barriers to development and designing programs that work across these barriers. Borders often

Table 8.1 Regional cooperation in the South China Sea

Acronym	*Initiative*
BIMP-EAGA	Brunei Darussalam-Indonesia-Malaysia-Philippines East ASEAN Growth Area
GMS	Greater Mekong Subregion
IMT-GT	Indonesia-Malaysia-Thailand Growth Triangle
PBG	Pan-Beibu Gulf Economic Cooperation (Gulf of Tonkin)
SIJORI/IMS-GT	Singapore-Johor-Riau Growth Triangle / Indonesia-Malaysia-Singapore Growth Triangle

South China Sea Membership	*Brunei*	*China*	*Indonesia*	*Malaysia*	*Philippines*	*Viet Nam*
BIMP-EAGA	X		X	X	X	
GMS[a]		X[b]				X
IMT-GT			X	X		
PBG[c]	X	X[d]	X	X	X	X
SIJORI/IMS-GT			X	X		

Notes: (a) Additional members: Cambodia, Lao PDR, Myanmar, and Thailand; (b) China participates in the GMS through Yunnan Province and the Guangxi Zhuang Autonomous Region; (c) every ASEAN member can participate in the Chinese-sponsored PBG; however the focus is on the archipelagic states and Viet Nam (ADB 2011, p. 1); and (d) in China, the PBG particularly involves Guangxi Zhuang Autonomous Region and Guangdong and Hainan provinces (ADB 2011, p. 1)

do not define areas that are optimal from the standpoint of economic development:

- Perhaps the borders divide a natural economic zone such as a river valley and the tariffs or other barriers to trade, reduce the scope for economies of scale on either side. Reducing barriers to cross-border businesses might encourage local investment and growth.
- Borders can delineate very different economies; Singapore differs from Indonesia in many more ways than not. Opening borders through regional cooperation may provide for complementary resource use, for instance, capital from a high-income country could be invested in a low-wage neighbor.
- Borderlands are often far from the centers of power and economy and as such may not receive public investment for infrastructure especially for transport. Opening the borders and collaboratively providing for infrastructure may act as catalysts to sustainable development, in turn providing for adequate return for the infrastructure investment.

Development tends to be sui generis, demanding individualized solutions, so regional economic initiatives have to be designed to meet specific opportunities and needs. A successful initiative in one part of the world cannot simply be replicated in another, but the lessons from one can be useful in planning for others.

In theory, ASEAN as an organization stands behind some of these different *subregional* initiatives, as they are sometimes referred to, especially BIMP-EAGA and IMT-GT. However, the initiatives listed in Table 8.1 operate almost completely independent of ASEAN as an organization. This could change and ASEAN could well play an important role in providing a third option for reducing South China Sea conflicts, but it would take considerable changes in operational mandates and resources, both financial and staff (ADBI 2014, esp. pp. 272–275).

8.1 The Greater Mekong Subregion

Perhaps the most pertinent regional cooperation experience is that of the GMS.[4] The GMS was founded in 1992 during a time of considerable turmoil: active conflict on the mainland of Southeast Asia was a recent memory, and political institutions at all levels in several countries were still quite fragile. Although lacking formal bureaucracy, with Asian Development Bank staff acting as a secretariat, the GMS has organized a wide range of collaborative efforts, notably through 2013 roughly $16.6 billion in development investment. These investments were mostly for transport infrastructure that has helped underpin an expansion of trade within this part of Southeast Asia.[5] Improving especially highways within the region, the GMS has underwritten a transformation of the ability to move goods and people between the economic centers of the different countries (Green 2007 and 2008b). One measure of the impact of the infrastructure investment was the growth of regional trade: "The size of intra-GMS trade increased from $13.9 billion in 2000 to $81.2 billion in 2009, for an average annual growth rate of 21.7 %" (Srivastava and Kumar 2012, p. 23).

The GMS could be considered a co-facilitator, alongside Factory Asia, of the growth and development seen in mainland Southeast Asia over the past two decades. The GMS was initiated at the time when China and Viet Nam were transforming their economies and supported their integration into the global economy, and at the time when firms were looking to find centers for manufacturing and assembly with transport links to the rest of Asia. By focusing on transport, especially regional roads, the GMS program provided a key element in encouraging trade and economic growth. Additionally, an important part of this has been to harmonize *soft infrastructure*—the policy counterpart of the hard infrastructure of roads and highways, railways, and airports. Wiemer (2009, p. 13) emphasizes the importance of the long running, if uneven, work on trade and transport facilitation (TTF), which has encompassed a wide range of efforts, including such practical steps as encouraging *single-stop* customs inspection.[6]

The GMS brings together two South China Sea claimants, China and Viet Nam. Moreover, the GMS was established in a region replete with unresolved transborder issues, some of which continue to threaten peace. In 2009, for instance, Cambodia and Thailand fell into small-scale armed conflict over disputed claims to an ancient temple area at their border (BBC 2013). In the simple world of the game spelled out above in Tables 7.1 and 7.2, there should be no resort to violence as the gains from collaboration dominated even successful aggression.

As suggested above, this cannot be a complete picture of interactions between two nations. The world is not simple and policymaking, especially with respect to the use of force at a country's borders, is sometimes determined by actions taken at the local level without considering broader national interests. However, it is significant that, in spite of armed clashes between Cambodian and Thai military forces at the border in mid-2009, resulting in fatalities on both sides, GMS meetings continued to be held.[7] At some meetings, the author observed the government officials working to make sure border conflicts did not interrupt their immediate work. In a larger case, although both countries are members of the GMS, neither China nor Viet Nam has completely eschewed aggressive behavior in defense of South China Sea claims. The mid-2014 contretemps between the two countries that resulted in deaths and some destruction to foreign-owned firms in Viet Nam was evidence of this. It is worth noting that by the end of the year, both these two countries could participate in the GMS Leaders' Summit in December in Bangkok with some observers noting that both governments appeared to use the meeting as a venue for building better relationships.[8]

Regional cooperation does not simply replace conflict with collaboration. Indeed, some of the interminable meetings in regional cooperation initiatives result from the intrinsic difficulties in dividing the costs and benefits from cross-border development projects. Regional cooperation brings its own sources of conflict. Investing, for example, in hydroelectric facilities in Laos to supply households and businesses in neighboring Thailand requires a host of problems to be resolved: What environmental rules are followed? How is social impact measured and mitigated? Where does the financing come from? What is the price of electricity and how should it change over time? That there are real net benefits, provides the scope for resolving these and other problems and for trading off against other cross-border issues, potentially including those that have security aspects. Thus, it might be easier to relax an aggressive stance toward conflicting cross-border territorial claims if there are clear benefits from other cross-border interactions.

Regional cooperation initiatives like the GMS also act to reduce the likelihood of military conflict by expanding policymakers' understanding of their cross-border counterparts and their shared interests. Dosch and Hensengerth (2005, p. 284) emphasize the role GMS activities play in that they "facilitate regular information exchange and better access to information for each of the actors, a precondition for growing trust and confidence."[9] The seemingly endless round of meetings that underpin regional cooperation efforts, in and of themselves, provide a context for less destructive forms of dispute resolution. The GMS has institutionalized this through the Phnom Penh Plan for Development Management, a large program that brings together government officials to study development problems from a regional standpoint. In roughly a decade since its beginning in 2003, nearly $11 million have been committed to train more than 2000 government officials in more than 100 programs building a sense of shared interests (ADB 2012).

It is, of course, impossible to finely delineate the influence of the GMS on the wider relationships between neighboring countries on the mainland of Southeast Asia. Countries in the Mekong have had long and sometimes contentious relationships. Their divergent histories, economic and political systems, and interests will naturally breed occasions for conflict. But increasing trade and cross-border investments and steps toward economic integration form the basis for national interests in cooperation, raising the costs of conflict, the benefits and efficacy of collaboration. As a visible part of this process, the GMS appears to function in a manner that encourages the participating countries in maintaining generally peaceful relations.

8.2 The Indonesia-Malaysia-Singapore—Growth Triangle

A second regional cooperation initiative, the Indonesia-Malaysia-Singapore—Growth Triangle (IMS-GT), also has been a catalyst for growth and to reducing cross-border strife, this time between Singapore and its neighbors Indonesia (with the growth triangle covering especially the Riau Islands, including Batam Island) and Malaysia (particularly the state of Johor). The initiative was established in 1989, initially as the Singapore-Johor-Riau (SIJORI) Growth Triangle: "The pact would combine the management expertise, rich capital, technology and infrastructure of Singapore with the abundant labour, land and natural resources of" the neighbors (Nor-Afidah 2002). The framework supports investment flows from Singapore and elsewhere, locating production facilities across the border in lower-wage regions, but near to Singapore's international logistics and financial industries.[10]

For the Malaysian side, the process appears to have been transformational with "a phenomenal period of growth for Johor which averaged GDP growth rates above 9 per cent through all of the 1990s up until the 1997 financial crisis."[11] The Indonesian region also showed tremendous growth. The Batam Indonesia Free Zone Authority (BIFZA), on a small (415 sq km) island 20 km from Singapore, at the center of Indonesian participation, recorded $17.7 billion in cumulative investments by 2014, supporting a workforce of nearly one-third million people.[12] As with the GMS, much of the hard effort of the IMS-GT came in the form of ensuring a policy environment that was supportive of cross-border trade and travel and of foreign investment—in particular, harmonizing standards and document requirements and ensuring a governance structure that gave confidence to investors, both foreign and domestic. Chia (1997, p. 300) noted, for instance, that Indonesia liberalized foreign investment regulations in Batam to encourage cross-border investment flows.

On the question of regional cooperation reducing conflicts, Bunnell et al. (2012, p. 470) warn that a "history of struggle continues to remain active and stirs just beneath the surface of the landscape." Land reclamation, transboundary haze, cross-border trade in sand and water, labor migration, and border delineation are all issues that divide these nations. As with the GMS, the IMS-GT does not provide the basis to eliminate disputes. Indeed, as with the GMS, the introduction of regional cooperation provides further room for conflict in fighting over the division of the benefits. For example, the location of manufacturing centers does not automatically provide for a division in the value-added generated—it is a subject for negotiation. But the possibility of trading over the division of benefits from regional cooperation allows for the possibility of bargaining over other issues, including those more directly related to security such as control of disputed territory. Regional cooperation does not automatically eliminate conflict, but it can put more chips on the table for each player and government officials readily affirm the value of the IMS-GT.[13]

There are significant differences between the two initiatives, including that of scale with the GMS covering all of mainland Southeast Asia and IMS-GT a few islands and a corner of the Malay Peninsula. Equally significant, while the GMS does not have a formal institutional structure, there is no headquarters; for instance, the ADB has provided continuous and considerable assistance as the secretariat and the work is anchored by a calendar-challenging set of meetings including a summit, every three years, of the participating leaders of all of the countries.[14] IMS-GT is more simply an effort defined largely through bilateral agreements and arrangements between the countries involved especially between Indonesia and Singapore, and between Malaysia and Singapore.

From the standpoint of impact, in retrospect, both initiatives benefited from contemporary economic trends, especially large wage differentials (in IMS-GT between Singapore on one side and Malaysia and Indonesia on the other; in the GMS between the OECD countries and less developed China and Southeast Asia). At the same time, changes in logistics (shipping containers) and communications (Internet) allowed for global value chains to be draped throughout East and Southeast Asia. These and other developments, including the opening of China and the former command-economies of mainland Southeast Asia to global markets, allowed the GMS and IMS-GT to have a real impact. The two initiatives capitalized on these trends by (i) the opening of the national borders to allow for the movement of capital and goods to exploit differences in wages and other factors such as land costs and (ii) the investment in infrastructure that lowered the costs of transport and manufacturing.

In contrast to the GMS and IMS-GT, the other regional cooperation initiatives listed in Table 8.1 have struggled to show substantial impact, providing a cautionary note on the efficacy of regional cooperation to reduce the risk of conflict in the South China Sea.

8.3 BIMP-EAGA and IMT-GT

On the southern flank of the South China Sea, the Brunei-Indonesia-Malaysia-Philippines—East ASEAN Growth Area (BIMP-EAGA) was put forward initially in 1992 by then Philippine President Fidel V. Ramos. He hoped to bring together the four countries to encourage development in the islands that had traditionally lagged the more vibrant national economic centers. In the Philippines, this was especially Mindanao, and for Indonesia and Malaysia, the island of Borneo. Similarly, the Indonesia-Malaysia-Thailand—Growth Triangle (IMT-GT) was established in 1993, growing to encompass Indonesia's Sumatra Island, the west coast and northern part of Peninsular Malaysia, and southern Thailand. Again these were areas that had lagged the development of the more dynamic parts of the three countries. In both cases, it was hoped that lowering cross-border trade and travel barriers, investing in trade facilitating infrastructure, and encouraging the private sector to work across borders would hasten growth.

In sharp contrast to the GMS and IMS-GT, neither BIMP-EAGA nor IMT-GT has generated either a large infrastructure program or strong private investment. The ADB has recently financed its first BIMP-EAGA regional project, a large transmission project to bring hydropower from Malaysian Borneo to the Indonesian part of the island (ADB 2013c). This type of project improves resource efficiency and income—it is a model regional cooperation project—but BIMP-EAGA has not generated public investment on the scale needed to transform the areas targeted. Similarly, while IMT-GT has identified local infrastructure needs, the program has been unable to significantly meet these needs (IMT-GT 2012, pp. 3–4, 9).

There has been somewhat more progress made on *soft infrastructure*, especially in those policy areas that improve connectivity—the ability to cross borders for trade, travel, or transit.[15] Both initiatives have provided for the private sector to reach across borders and to start building a regional business community, one that can gradually make better use of local resources and the changing global economic environment.[16] This is consistent with a broader propensity that Frost (2008, pp. 15–16) identifies for Asian policymakers of valuing "community building" in and of itself as a part of a long-term integrating process. These successes, useful as they are, have not been able to materially quicken economic development.

There are several reasons why neither BIMP-EAGA nor IMT-GT have played the role that the GMS or IMS-GT have in their respective regions. As a number of observers have suggested, both BIMP-EAGA and IMT-GT bring together relatively poor areas with more commonality in resource endowments than complementarity (Chia 1997, p. 301). Formally, rich-Brunei anchors BIMP-EAGA as Singapore does IMS-GT, but Singapore is much larger (Table 8.2). As a small oil and gas exporting country Brunei's private sector is proportionally smaller than Singapore's and cannot provide the needed investment in a much larger geographic area: EAGA covers three time zones, IMS-GT a relatively tiny area at the tip of the Malay Peninsula. IMT-GT is at more of a disadvantage in this respect, lacking a high-income anchor for cross-border investment.

The GMS was able to mobilize infrastructure investment that encouraged foreign investment from outside Southeast Asia, investments that linked production and trade to China; the IMS-GT similarly channeled funds especially from Singapore. While Indonesia, Malaysia, Thailand and the Philippines have all received inflows of foreign capital; these have largely not been directed toward the EAGA or IMT-GT regions. Those that have were not sufficient to change the nature of development in the target regions.

Part of the larger problem relates to the lack of infrastructure and the high costs of logistics in the program areas relative to what is offered to potential investors elsewhere, in coastal China, metropolitan Bangkok, or the southern part of Peninsular Malaysia. Inefficient ports, lack of airline connections, and poor roadways result in limited connectivity and have discouraged both domestic and foreign investment in EAGA and IMT-GT. The lack of transport and travel connections has been well recognized by the BIMP-EAGA and IMT-GT programs.[17] Indeed, this is precisely one of the reasons Philippine President Ramos proposed BIMP-EAGA: historically, providing reliable, economically viable connectivity between the roughly 24,000 islands in the archipelagoes of Indonesia

Table 8.2 The regional cooperation anchors

	GDP/capita	*GDP (constant 2005 $ Billion)*
Brunei	40,980	9.9
Singapore	56,284	208.3

Notes: http://data.worldbank.org/indicator

and the Philippines has been an unmet challenge.[18] While the poor island communities were often linked by small boats working in the shadow or informal economy, these did not allow for businesses to safely and predictably move people or cargo. Moreover, the unsettled security issues with piracy and smuggling ever present also discouraged investment.[19]

Meeting the needs of businesses for connectivity in the GMS partly involved building and upgrading the highway system. In the IMS-GT it similarly demanded better infrastructure systems between Singapore and its close neighbors, including dealing with the relatively short stretch of ocean to Batam Island. Conversely with BIMP-EAGA, while some of the challenges lie on land in Borneo, truly integrating these diverse borderlands would require connecting through economically viable air and sea links, Philippine ports on Mindanao with Indonesian and Malaysian ports on Borneo, a distance of hundreds of miles. With IMT-GT, the narrower but still substantial Strait of Malacca is a barrier to trade and travel.

As with other areas of transport, such as the use of containers, technological and regulatory change may offer scope for broadened trade and development in island geographies. Most relevantly, the Philippines has experimented with small-scale roll-on/roll-off (Ro-Ro) ferry systems to link together the country's diverse islands. In contrast to traditional sea transport systems where cargo is loaded and unloaded on relatively large freight vessels, Ro-Ro utilizes the shippers own vehicles: "the cargoes are rolling cargoes—i.e., cars, buses, trucks, etc.—which simply roll on and off the Ro-Ro ships, [so] Ro-Ro does not require cargo handling services" (ADB 2010, p. 4). The Philippine Ro-Ro system uses relatively small vessels along several national nautical *highways* running the length of the archipelago. Little more than one decade old, the combined investment and regulatory changes have resulted in lower shipping costs, from between 20–68 % in some instances, along with more frequent shipping schedules (ADB 2010, p. 15). A number of localities have seen dramatic increases in trade and tourism stimulating investment in local businesses such as logistics.

ASEAN has committed to building on this experience and establishing Ro-Ro links between port-pairs in Southeast Asia.[20] In theory, this sort of system could dramatically boost the economies of poorer islands, especially allowing small businesses to grow. In practice, in addition to mobilizing capital for new port facilities and vessels, there are a host of regulatory issues that need to be addressed related to recognizing foreign registration and insurance for vehicles that cross national borders. Moreover, the countries differ from one another on some basic transport attributes. In the

Philippines, for example, vehicles drive on the right side of the road (using left-hand-drive vehicles), in Indonesia and Malaysia, the opposite is the case. In the Philippines, it is illegal to use a right-hand-drive vehicle, severely limiting the efficacy of an international Ro-Ro (GMA News Online 2011).

In summary, neither BIMP-EAGA nor IMT-GT has had a sizeable impact on the local economies and, for all the worthwhile planning and discussion, face serious problems in becoming more successful. Turning back to the issue of the South China Sea disputes, we need to ask if either BIMP-EAGA or IMT-GT have contributed to conflict-alternative paths to resolving disputes. Formally, the countries consistently and regularly affirm support for these initiatives.[21] They devote considerable resources to the consultative processes, the stream of meetings that underlie these initiatives. However, given that these programs have not established large benefits this argument should not be pressed too far.

The recent flaring up of the dispute over Sabah between Malaysia and the Philippines offers some cautionary insights. The Sabah dispute has formal similarities to the South China Sea conflicts: territory occupied by one country, claimed by another; periodic conflict; and aggressive public posturing and appeals to nationalistic sentiment (Box 8.1). Indeed, no less an authority than former Philippines President Fidel Ramos—the originator of the BIMP-EAGA program—stated in 1994 that the program depended on being able "to set aside the contentious issue of [the Philippines' claim to part of] Sabah and allow the expansion of economic relations" (Ramos 1994). Unfortunately, since that time, unlike what has been seen in the GMS and IMS-GT, likely because there are fewer clear benefits from BIMP-EAGA, the Philippine Government has been willing to slow down some aspects of the initiative's operations in defense of its territorial claims.[22] Neither IMT-GT nor BIMP-EAGA generated the kind of impact that greatly alters the basic relationships between the countries involved.[23]

Box 8.1 The Issue of Sovereignty over Sabah A historic claim to the Malaysian state of Sabah, on northeastern Borneo, by the Sultan of Sulu has sometimes embroiled the governments of Malaysia and the Philippines. The dispute dates to the late nineteenth century when the British North Borneo Company and the Sultanate of Sulu signed an agreement concerning the control of Sabah. Depending on the version of this treaty, Sabah was either ceded to British control or simply leased.

An annual payment by the successor to the British, the Malaysian Government, to the successors of the Sultan of Sulu has done nothing to resolve the issue. While the Philippine Government has rarely taken definitive steps to advance any claim, it has equally been reluctant to publicly renounce them. Thus, in 2013, when a group of armed Filipinos made their way to Sabah to seek recognition of the claims by one of the pretenders to the Sultanate of Sulu, it was not simply a security issue for Malaysia, but an international incident (Mullen 2013).

8.4 Pan-Beibu Gulf Economic Cooperation

Centered in the South China Sea, involving all ASEAN members, the Pan-Beibu Gulf (Gulf of Tonkin) initiative is intended to enhance ASEAN-Chinese economic relations. It has wide ambitions:

> (i) expanding infrastructure connectivity beyond the GMS to link the rest of ASEAN with the PRC; (ii) developing ports and maritime transport services; (iii) facilitating cross-border trade and investment with a focus on developing trade and economic cooperation zones; (iv) strengthening agriculture cooperation, especially with a view to solving food security and food safety issues; and (v) promoting industrial cooperation and developing regional value chains.[24]

In practice, the Chinese initiative (i) brings together ASEAN and Chinese business representatives for promotional meetings and (ii) encourages development in Guangxi Zhuang Autonomous Region.[25] As such it could be compared with BIMP-EAGA and IMT-GT in providing business networking and highlighting infrastructure and policy needs. It lacks the organizational capacity to initiate infrastructure projects in ASEAN members or to negotiate policy changes.[26]Thus, like BIMP-EAGA and IMT-GT, it provides for community building among the private sector but does not materially provide a modality to reduce the temptation to settle South China Sea disputes through conflict. Conversely, continued tension in the South China Sea, particularly between China and Viet Nam, has likely impeded any attempt to strengthen the Pan-Beibu initiative (Hosokawa 2009, especially p. 73). Similarly, these disputes may hamper the very ambitious Chinese proposal for a Maritime Silk Road, suggested in October 2013 by Chinese President Xi Jinping in a visit to Indonesia, an effort meant to encourage trade among the maritime ASEAN countries and China (Zheng 2014).

Notes

1. On a broader canvas, that of Asia, Frost (2008) examines how "regionalism" is fueled by market forces and deliberate government direction affecting both economic performance and national security.
2. APEC is the Asia-Pacific Economic Cooperation forum established in 1989. (http://www.apec.org/) ZOPFAN is the Zone of Peace, Freedom and Neutrality declaration, signed in 1971, helping to set a tone for political relations within and external to the region (Djalal 2011).
3. For a similar list of such initiatives, covering a broader geographic area, see Dosch and Hensengerth (2005, p. 271, Table 1). The authors also examine whether regional cooperation can encourage "stable and peaceful subregional relations" especially within the context of mainland Southeast Asia (p. 264).
4. General information on the GMS can be found at the Asian Development Bank-maintained web site: http://www.adb.org/countries/gms/main
5. Approximately $11 billion has supported infrastructure, largely in transportation (http://www.adb.org/countries/gms/sector-activities).
6. ADB (2015, pp. 8–10). The centerpiece of the GMS is the huge transport infrastructure program, but it has also supported a wide range of projects broadly related to development. A recent GMS publication, for instance, noted projects in agriculture, energy, environment, human resource development, tourism, transport and trade facilitation, and urban development (ADB 2015). A picture of the rich tapestry of a region in transformation can be found at ADB (2009a).
7. The record of GMS dialogue and a sense of the ongoing set of GMS meetings are found at ADB (2009b).
8. Kyozuka (2014) reports that "Although Vietnam and China are locked in a dispute over territory in the South China Sea, Vietnamese Prime Minister Nguyen Tan Dung offered little in the way of criticism of its neighbor at the summit."
9. The conclusion of Dosch and Hensengerth is worth reproducing: " the value of the GMS regarding economics and traditional security is not to be seen in direct interference through independent influence, which the institution is unable to exert, but rather in the particular importance each of the actors attaches to it, which is why diplomacy has largely replaced military endeavours" (2005, p. 284).
10. "For businesses located in Batam-Bintan and Johor, doing business with and through Singapore is much more cost efficient and time saving as Singapore provides world class transportation, telecommunications, financial, and commercial infrastructures" Chia (1997, p. 300).
11. Bunnell et al. (2012, p. 468). This compares to 7.1 % for the country as a whole. (Average annual rate for 1990–1999, for real GDP, from the World

Bank, World Development Indicators.) For a different take on SIJORI see Wiemer (2009), who emphasizes, among other issues the lack of institutional development and of "grassroots involvement" (pp. 13).

12. BIFZA. Bunnell, et al. discuss the impact of IMS-GT on the Riau Islands (2012, pp. 468–469).
13. Author's observations, in discussion with Singaporean and Batam officials, 2006–2010.
14. The interested reader is directed to see the list of meetings; many at the ministerial level, noted in the ADB published GMS e-updates (ADB 2015).
15. The document BIMP (2012, especially pp. 14–16) discusses work on customs, immigration, quarantine, and security (CIQS) issues.
16. BIMP-EAGA has established the BIMP-EAGA Business Council (BEBC, see https://bimpeagabc.com/?page_id=23) and IMT-GT the analogous Joint Business Council (JBC, see http://www.imtgt.org/Private-Sector.htm) to provide forums for private sector interests to meet and ensure their views are heard in the respective initiative's debates.
17. For BIMP-EAGA, in the Implementation Blueprint 2012–2016, the first strategic "pillar" or guiding principle for planning is Enhanced Connectivity (BIMP 2012, p. 5). Just as strongly, the analogous IMT-GT document states that "Connectivity, in its broad sense of linking geographic areas, facilitating economic transactions, and enhancing people-to-people interface, shall be the overarching objective" (IMT-GT 2012, p. 5).
18. "Physical isolation is the most pressing development challenge that these islands are facing" (ADB 2010, p. vi).
19. "With the exception of 2007 to 2012, when piracy in East Africa experienced a sharp increase, the South China Sea has been the most piracy-prone region in the world, with up to 150 attacks per year" Schoenberger (2014).
20. See, for example, the Joint Ministerial Statement of the Seventeenth ASEAN Transport Ministers Meeting Phnom Penh, Cambodia, 16 December 2011, section 11. Available http://www.asean.org/?static_post=joint-ministerial-statement-seventeenth-asean-transport-ministers-atm-meeting-phnom-penh-cambodia-16-december-2011 [Accessed Aug. 14, 2015].
21. See, for instance, the statements from the Leaders' Summits. The statement of the 10th BIMP-EAGA Summit in 2014 is available at http://www.asean.org/storage/images/documents/24thASEANSummit/Joint%20Statement%20of%20the%2010th%20BIMP-EAGA%20Summit.pdf. Typically the Leaders of the participating countries express their "satisfaction with the progress" in a variety of areas. The corresponding 8th IMT-GT Summit is available at http://www.asean.org/storage/images/documents/24thASEANSummit/The%208th%20IMT-GT%20Summit.pdf
22. In mid-2013, ADB Staff explained to the author that the Philippine Government had been reluctant to hasten the establishment of a Facilitation

Center for BIMP-EAGA in the Malaysian state of Sabah to avoid looking like it was abandoning claims to the territory.

23. As one further, if small, indication of this, unlike the GMS, although held annually, the IMT-GT and BIMP-EAGA leaders' summits are held alongside the ASEAN meetings and often non-substantive (author's observation).
24. ADB (2011, p. 2) referring to the conclusions of the "Feasibility Report on PBG Economic Cooperation, which was endorsed by the ASEAN–PRC Economic Ministers' Meeting on 12 August 2011."
25. The Pan-Beibu initiative is well integrated into Chinese development planning (Hosokawa 2009).
26. The author was a member of the International Expert Group of the Pan-Beibu Gulf Economic Cooperation Forum, 2008–2009. Hosokawa (2009, p. 74) notes the difficulty in arranging financing for infrastructure in Viet Nam.

Bibliography

Asian Development Bank (ADB). 2009a. Corridor chronicles: Profiles of cross border activities in the Greater Mekong subregion. http://adb.org/sites/default/files/pub/2009/corridor-chronicles-gms.pdf. Accessed 23 June 2015.

———. 2009b. REG: Greater Mekong subregion economic cooperation program: Joint ministerial statement: 15th ministerial conference, June 2009. http://www.adb.org/sites/default/files/page/42450/15th-summit-joint-ministerial-conference-statement-greater-mekong-subregion-gms.pdf. Accessed 23 June 2015.

———. 2010. Bridges across oceans: Initial impact assessment of the Philippines Nautical Highway System and lessons for Southeast Asia. [online] Publication, Apr 2010. http://www.adb.org/publications/bridges-across-oceans-initial-impact-assessment-philippines-nautical-highway-system-and. Accessed 14 Aug 2015.

———. 2011. Support for Pan-Beibu Gulf Economic Cooperation. Technical assistance report, Project number: 45002, Dec 2011. http://www.adb.org/sites/default/files/project-document/60295/45002-001-reg-tar.pdf. Accessed 22 June 2015.

———. 2012. Project details: 46232–001: Greater Mekong subregion Phnom Penh Plan for Development Management Phase V. http://www.adb.org/projects/46232-001/main. Accessed 6 Apr 2016.

———. 2013c. ADB's first BIMP-EAGA project to bring clean energy to West Kalimantan. 28 Aug 2013. News Release. http://www.adb.org/news/adbs-first-bimp-eaga-project-bring-clean-energy-west-kalimantan. Accessed 26 June 2015.

———. 2015. GMS e-Updates, September 2014–February 2015.| Vol. 8, Issue No. 2. http://www.adb.org/sites/default/files/publication/154923/gms-e-updates-sep2014-feb2015.pdf. Accessed 18 June 2015.

Asian Development Bank Institute (ADBI). 2014. ASEAN 2030: Toward a borderless economic community. Tokyo, Japan. http://adb.org/sites/default/files/pub/2014/asean-2030-toward-borderless-economic-community.pdf. Accessed 24 Aug 2015.

British Broadcasting Corporation (BBC). 2013. Q&A: Thailand-Cambodia temple dispute. BBC News, 7 Nov. 2013. http://www.bbc.com/news/world-asia-pacific-12378001. Accessed 23 June 2015.

Brunei Darussalam-Indonesia-Malaysia-Philippines–East ASEAN Growth Area (BIMP). 2012. Implementation blueprint 2012–2016, Mar 2012. http://www.adb.org/sites/default/files/page/34232/bimp-eaga-implementation-blueprint-2012-2016.pdf. Accessed 6 Apr 2016.

Bunnell, Tim, Carl Grundy-Warr, James D. Sidaway, and Matthew Sparke. 2012. Geographies of power in the Indonesia-Malaysia-Singapore growth triangle. In *International handbook of globalization and world cities*, ed. B. Derudder, M. Hoyler, P.J. Taylor, and F. Witlox, 465–475. Cheltenham/Northampton: Edward Elgar. http://faculty.washington.edu/sparke/geopower.pdf. Accessed 18 June 2015.

Chia, Siow Yue. 1997. Regionalism and subregionalism in ASEAN: The free trade area and growth triangle models. In *Regionalism versus multilateral trade arrangements*, NBER-EASE Volume 6, ed. Takatoshi Ito and Anne O. Krueger, 275–312. University of Chicago Press. http://www.nber.org/chapters/c8603. Accessed 6 Apr 2016.

Djalal, Hasjim. 2011. Rethinking the zone of peace, freedom and neutrality (ZOPFAN) in the Post-Cold War era. 25th Asia-Pacific Roundtable, 31 May 2011. http://www.isis.org.my/files/25APR/paper_ps_5_hasjim_djalal.pdf. Accessed 17 Jun 2015.

Dosch, Jörn, and Oliver Hensengerth. 2005. Subregional cooperation in Southeast Asia: The Mekong Basin. *European Journal of East Asian Studies* 4(2): 263–286. http://www.wiwi.uni-rostock.de/fileadmin/Institute/IPV/Lehrstuehle/Internationale_Politik/J%C3%B6rn_Dosch/Dosch_Hensengerth.pdf. Accessed 23 Jun 2015.

Frost, Ellen L. 2008. *Asia's new regionalism*. Boulder: Lynne Rienner.

GMA News Online. 2011. Narrower 'right-hand' drive ban to run over truck, bus imports? Money. http://www.gmanetwork.com/news/story/210733/money/narrower-right-hand-drive-ban-to-run-over-truck-bus-imports. Accessed 22 Feb 2016.

Green, David Jay. 2007. Bridging the ASEAN developmental divide: Challenges and prospects, a regional overview. *ASEAN Economic Bulletin* April(24/#1): 15–34.

———. 2008a. Alphabet soup and the ASEAN neighborhoods. *The Jakarta Post*, 4 Jan 2008. http://www.adb.org/news/op-ed/alphabet-soup-and-asean-neighborhoods. Accessed 15 June 2015.

———. 2008b. The role of ASEAN economic community as a commitment to policy certainty. *ASEAN Economic Bulletin* August(25/#2): 209–227.

Hosokawa, Daisuke. 2009. Pan-Beibu Gulf Economic Cooperation—China's new initiative in cooperation with ASEAN. *Osaka Keidai Ronshu* 60(2 July): 67–78. www.osaka-ue.ac.jp/file/general/4626. Accessed 11 Aug 2015.

Indonesia-Malaysia-Thailand–Growth Triangle (IMT-GT). 2012. Implementation blueprint 2012–2016. http://imtgt.org/Documents/IMT-GT_Implementation_Blueprint_2012-2016.pdf. Accessed 27 Jun 2015.

Kyozuka, Tamaki. 2014. China's influence campaign: Beijing pushes 'go south' policy at Greater Mekong summit. Nikkei Asian Review. Asia.nikkei.com, 20 December 2014. http://asia.nikkei.com/print/article/66589. Accessed 5 Jul. 2015.

Mullen, Jethro. 2013. Malaysia launches attack on Filipino intruders in Borneo. CNN, 5 Mar 2013. http://www.cnn.com/2013/03/04/world/asia/malaysia-philippines-standoff/. Accessed 3 Jul 2015.

Nor-Afidah. 2002. Growth triangle | Infopedia. National Library Board Singapore. http://eresources.nlb.gov.sg/infopedia/articles/SIP_58_2005-01-06.html?utm_expid=85360850-6.qNOOYF40RhKK6gXsQEaAJA.0&utm_referrer=https%3A%2F%2Fwww.google.com%2F. Accessed 18 Jun 2015.

Ramos, Fidel. 1994. Cross-border growth triangles promote prosperity in East Asia. International Herald Tribune. 30 Nov 1994. Nytimes.Com. http://www.nytimes.com/1994/11/30/opinion/30iht-edramos.html. Accessed 16 Aug 2015.

Schoenberger, Sarah. 2014. Piracy in the South China Sea: Petty theft in Indonesia, kidnapped ships in Malaysia. Center for International Maritime Security, 6 Sept 2014. http://cimsec.org/piracy-south-china-sea-petty-theft-indonesia-kidnapped-ships-malaysia/12899. Accessed 18 May 2015.

Srivastava, Pradeep, and Utsav Kumar. 2012. GMS trade—trends and patterns. Chapter 2 (pp. 11–37). In *Trade and trade facilitation in the Greater Mekong Subregion*, ed. Srivastava, Pradeep and Utsav Kumar. Mandaluyong City: Asian Development Bank. http://www.adb.org/sites/default/files/publication/29977/trade-and-trade-facilitation-gms.pdf. Accessed 17 June 2015.

Wiemer, Calla. 2009. Economic corridors for the Greater Mekong Subregion. East Asian Institute, Singapore, EAI background brief No. 479, 15 Sept 2009. http://www.eai.nus.edu.sg/publications/files/BB479.pdf. Accessed 6 Apr 2016.

Zheng, Shibo. 2014. Interview: ASEAN welcomes China's new Maritime Silk Road initiatives—Guangxi Beibu Gulf Economic Zone. Xinhuanet, 15 Aug 2014. http://news.xinhuanet.com/english/china/2014-08/15/c_133559198.htm. Accessed 2 Oct 2014.

9

South China Sea Regional Cooperation: A Tentative Exercise

Abstract This chapter argues that there indeed could be a Third Option for the South China Sea, regional collaboration can lead to quickened economic growth and development, but this is not simple to find or initiate. A list of eight guidelines is provided, summarizing the lessons from existing initiatives. The conclusion thus poses a challenge: while the effort needed to successfully mount a viable regional cooperation program is formidable, the risks to the regional and indeed the global economy are too large to allow the present behavior to go unchanged.

Keywords Regional cooperation • South China Sea • Conflict

What would be the nature of a regional cooperation initiative that could provide an alternative to conflict in the South China Sea? It would have to be widely seen as providing benefits, especially in trade and economic development, that could be used to offset the foregone, if uncertain, gains from aggressive behavior in the South China Sea. The discussion above suggests real challenges to effectively structuring such a regional cooperation initiative. In trying to outline such an effort, we leave the firm contours of the South China Sea for Terra Incognita; at best, the section below puts forward some underlying principles as a starting point for further work. Table 9.1 summarizes the recommendations.

D.J. Green, *The Third Option for the South China Sea*,
DOI 10.1007/978-3-319-40274-1_9

Table 9.1 Principal recommendations—South China Sea regional cooperation

1.	Focus on providing shared benefits between China and the Philippines and between China and Viet Nam
2.	Provide clear and substantial economic benefits by raising the potential for trade and cross-border investment
3.	While trade promotion and growth enhancement are the highest priority, the goals of the initiative cannot be limited to this
4.	Ensure that the program results in infrastructure investment—infrastructure that improves connectivity
5.	Ensure that business investment will flow
6.	Hold in abeyance territorial claims to the seas and the resources
7.	Jointly harvest, in a sustainable fashion, the resources, utilizing the revenues for common development, especially in coastal areas
8.	Obtain multilateral support for administering the regional project

1. Focus on providing shared benefits between China and the Philippines and between China and Viet Nam.

These pairs of countries are currently experiencing the highest level of tension; this is where the alternative policy choices are most needed. Without resolving the issues between these pairs of countries, confrontations will continue to take place; business investment, trade, and tourism interrupted; and ASEAN–Chinese relations held hostage. Without specifically calming Chinese–Philippines relations, there will always be calls for the USA to become more involved. Even if these will not be acted upon, they can continue to roil the waters. This is not to say, "ignore the other disputes." Indeed, if a program can be found to calm the more contentious claimants, it is likely to ease worries by the other countries, Brunei, Indonesia, and Malaysia.

2. Provide clear and substantial economic benefits by raising the potential for trade and cross-border investment.

A regional cooperation initiative to calm international relations in the South China Sea would convincingly need to provide broad economic gains—gains that would be sufficient to allow political leaders to walk back from their public commitments to capture the South China Sea for exclusive exploitation. This would likely mean activities that would encourage Chinese investment in Indonesia, the Philippines, and Viet Nam, investment that would stimulate trade and regional growth on both sides of the South China Sea.

Between China and Viet Nam, this could mean strengthening existing GMS programs, as this framework already supports the kind of effort needed. To meaningfully change the existing relationship would take a high-level commitment on the part of both countries, one that explicitly seeks alternatives to the present focus on security and nationalistic sentiment. It would take a cold calculation and frank public admission that more investment, trade, employment, and income are worth sacrificing an aggressive South China Sea posture. This action would build on existing trends—Chinese outward FDI is increasing with Viet Nam being already identified as a useful target and China and Viet Nam are significant trading partners. Thus, this is less a speculative suggestion than a matter of weighing costs and benefits and finding practical ways of enhancing cross-border trade and capital flows.

If the nature of the regional cooperation initiative between China and Viet Nam is relatively straightforward, the needed programs between China and the Philippines and between China and Indonesia are less clear. In the abstract, however, the poorer islands in the central and south Philippines and in Indonesian Borneo and Sulawesi provide targets for natural resource-based Chinese FDI. Addressing connectivity issues, especially infrastructure, and providing investment funds could unlock potentially rich sources of agricultural and marine products to meet growing Chinese demand and meet long-running needs for employment and livelihood in the poorer islands of Indonesia and the Philippines. Baldly stated this is unarguable; however, the lack of success of BIMP-EAGA and IMT-GT warns that this is by no means easy to accomplish. But the recent experience of the Philippine Ro-Ro system suggests that the historical problems of connectivity may not be so intractable. Indonesia's experience with the IMS-GT provides some useful policy precedents. Moreover, the gradual, if uneven, establishment of the ASEAN Economic Community is pushing all Southeast Asian countries to adopt policies that allow for more openness, policies that would be more welcoming to foreign investment. Thus, there are existing or emerging economic trends that can be the foundation of a regional cooperation initiative that could calm the South China Seas.

3. While trade promotion and growth enhancement are the highest priority, the goals of the initiative cannot be limited to this.

The turmoil in the South China Sea has locked countries into defensive positions, into policies that identify nationalism and national security with territorial claims. Changing this will take more than the promise of the provision of faster regional growth; it will take appealing to interests that rise above narrow economic considerations. This could include the following:

- Climate change mitigation efforts
- Control of piracy, smuggling, and other illegal activities on the seas
- Coastal resource husbandry including control of pollution, sustainable aquaculture, and protection of reefs and mangroves
- People-to-people, community building including disseminating development lessons learned by local government leaders, business networking, and joint cultural programs

This point needs to be qualified: The lessons of other regional cooperation initiatives are reasonably clear, it is important to rise above the focus on growth, but if there is no observable impact on trade and local growth, the initiative will fail to have much weight in national decision-making.

4. Ensure that the program results in infrastructure investment—infrastructure that improves connectivity

There is little that can productively focus the attention of government officials as a large infrastructure program, one that can uncover economic potential. The lack of infrastructure investment has likely stymied growth in BIMP-EAGA and IMT-GT localities; conversely, GMS and IMS-GT have been successful partly through adding transport and logistics infrastructure. In both cases, but especially in the GMS, infrastructure that improves connectivity—the ability to move goods, services, and people across national borders—has been economically transformational. But infrastructure not only allows for trade and investment, it also draws the attention of government officials to soft policy lacunae if only in an effort to ensure that the expense of the infrastructure investment provides a commensurate return. In the South China Sea this would likely mean investment in port facilities to better connect southern China with archipelagic Southeast Asia and intensifying the GMS work at the border between China and Viet Nam. Ensuring that there is private sector involvement in infrastructure provision, through public–private partnerships, would reduce the overall resource costs, but also pave the way for the next needed element.

5. Ensure that business investment will flow.

For regional cooperation to yield significant benefits, to be part of the policymaking calculus, there must be trade-generating private sector investment. The Chinese Special Export Zones (SEZs) are well known and provide one useful model: foreign investment, responding to infrastructure and special legal and regulatory provisions, created assembly plants yielding trade and employment that helped power China's unprecedented growth over the last few decades. Indonesia, the Philippines, and Viet Nam have similar useful experiences. Indonesia's success in establishing production centers in Batam off Singapore was an important element underpinning IMS-GT. In the Philippines, there is, for example, the very successful Mactan Export Processing Zone on the island of Cebu. Vietnam has created a wide range of legal entities, including export processing zones and high-tech zones, to encourage investment, production, and trade. (Huyen and Nguyen 2013) While most of these are geared toward manufacturing, it is likely that regional cooperation could provide for increased agricultural or aquaculture exports from Indonesia, the Philippines, and Viet Nam to China. The key is to look for resource complementarities, or for natural economic zones where policy changes and infrastructure investment can encourage foreign and domestic businesses to invest.

6. Hold in abeyance territorial claims to the seas and the resources.
7. Jointly harvest, in a sustainable fashion, the resources, utilizing the revenues for common development, especially in coastal areas.

Many observers have suggested these elements in proposing resolutions to the South China Sea tensions.[1] In the region, Bernard (2013, p. 4) notes examples of such agreements including between Malaysia and Thailand and the Republic of Korea and China.[2]

The Chinese Ministry of Foreign Affairs maintains a web page that notes that more than three decades ago, Deng Xiaoping called for "setting aside dispute and pursuing joint development" (MOFA 2016). The Government's formal position is worth reporting:

> 1. The sovereignty of the territories concerned belongs to China. 2. When conditions are not ripe to bring about a thorough solution to territorial dispute, discussion on the issue of sovereignty may be postponed so that the dispute is set aside. To set aside dispute does not mean giving up sovereignty.

It is just to leave the dispute aside for the time being. 3. The territories under dispute may be developed in a joint way. 4. The purpose of joint development is to enhance mutual understanding through cooperation and create conditions for the eventual resolution of territorial ownership.

Thus, in China at least, the political cover, the public policy position, allowing for cooperative resource exploitation alongside unresolved territorial claims has a longstanding and enviable pedigree. As argued above, this theoretical policy has been overshadowed by the rude calculus that aggressive behavior would bring larger benefits: the absence of a regional cooperation alternative that clearly provides benefits in excess of that from any other behavior has led to periodic confrontations risking a slide to full conflict.

8. Obtain multilateral support for administering the regional project.

ADB's experience in supporting GMS is very pertinent: the initiative provided for true cooperation in very difficult circumstances for two of the same countries now locked in a no-win stalemate in the South China Sea.[3] Conversely, the Pan-Beibu exercise suggests how difficult it is for any one country to take the lead and administer an activity that needs to be seen as equitable in the sharing of costs and benefits.

As Lord Palmerston noted, we cannot depend on the friendship between countries; countries will look to their interests.[4]The task then is to create the interests—the shared benefits—that make cooperation, rather than conflict worthwhile.

Notes

1. See, for instance, Rogers (2012, p. 94) and Khemakorn (2006).
2. See also US EIA (2013): "Malaysia and Thailand agreed to develop a section of the Gulf of Thailand jointly without either party ceding legal rights to it."
3. A similar experience can be seen in the Central Asia Regional Economic Cooperation (CAREC) effort that originally brought together China (focusing on the Xinjiang Uygur Autonomous Region), Kazakhstan, the Kyrgyz Republic, and Uzbekistan. The ADB initiated the effort in the mid-1990s, bringing together countries that had only recently become independent, lacked government administrative capacity, and as market-based institutions were struggling to establish themselves in the former Soviet

Union. This cooperative exercise depended completely on ADB resources. See the web site: http://www.carecprogram.org/

4. "We have no eternal allies, and we have no perpetual enemies. Our interests are eternal and perpetual, and those interests it is our duty to follow." Henry John Temple Palmerston, Remarks in the House of Commons, March 1, 1848 (Palmerston 1848).

Bibliography

Bernard, Leonardo. 2013. Prospect for joint development in the South China Sea. "Managing tensions in the South China Sea" conference held by CSIS on June 5–6, 2013. http://csis.org/files/attachments/130606_Bernard_Conference-Paper.pdf. Accessed 22 Nov 2014.

Huyen, Hoang Thu, and My Nguyen. 2013. Development zones in Vietnam. Vietnam Briefing News, 4 Jun 2013. http://www.vietnam-briefing.com/news/development-zones-vietnam.html/. Accessed 7 Apr 2016.

Khemakorn, Pakjuta. 2006. Sustainable management of pelagic fisheries in the South China Sea Region, Nov 2006. http://www.un.org/depts/los/nippon/unnff_programme_home/fellows_pages/fellows_papers/khemakorn_0607_thailand.pdf. Accessed 17 May 2015.

Ministry of Foreign Affairs. People's Republic of China (MOFA, China). 2016. Set aside dispute and pursue joint development. http://www.fmprc.gov.cn/mfa_eng/ziliao_665539/3602_665543/3604_665547/t18023.shtml. Accessed 7 Apr 2016.

Palmerston, Henry John Temple. 1848. Remarks in the house of commons, the quotations page. http://www.quotationspage.com/quote/41290.html. Accessed 17 Feb 2016.

Rogers, Will. 2012. The role of natural resources in the South China Sea. In *Cooperation from strength the United States, China and the South China Sea*, ed. P. Cronin, 83–97. Washington, DC: Center for a New American Security. http://www.cnas.org/sites/default/files/publications-pdf/CNAS. Accessed 16 May 2015.

United States Energy Information Administration (US EIA). 2013. South China Sea. http://www.eia.gov/beta/international/regions-topics.cfm?RegionTopicID=SCS. Accessed 23 Jul 2015.

10

Conclusion

Abstract This concluding chapter reprises the themes of the book. In the South China Sea, disputed claims to ownership of the small islands, the control of passage and the energy and fishery resources are growing sources of tension. A simple game theory model helps explain behavior, warning that there are risks of real conflict—conflict that could derail regional development. Regional economic cooperation can be an alternative to the present conflicts. However, while there is the potential for peaceful development of the South China Sea, there are real challenges to structuring successful programs.

Keywords ASEAN • China • Game theory • Philippines • Viet Nam • Regional cooperation

For decades, countries bordering the South China Sea have been engaged in campaigns to claim title to the maritime territory and the resources therein. Passive advocacy of claims has alternated with more aggressive behavior—resulting in some cases in the loss of life and repercussions for the broader economies. The book, using a simple game theory model, suggests that the countries involved in these disputes, behave as if the cost-to-benefit ratio associated with aggressive behavior is low. This may have been true in the past, but with the progressive establishment of Factory Asia, with China and Viet Nam becoming more integrated into

D.J. Green, *The Third Option for the South China Sea*,
DOI 10.1007/978-3-319-40274-1_10

the global economy, tensions in the South China Sea present real risks of disruption to the trade and capital flows that have underwritten East and Southeast Asian growth and development. Against this, the possible economic rewards of exploiting energy and fishery resources pale in comparison—there is little in the nature of real treasure that would entice a rational player to risk economy-wide shocks.

There are other explanations as to why governments have continued to tussle over the South China Sea: the national security card is easy to play in this situation and difficult to trump by appealing to economic concerns. And national security is only one of the more than one-dozen other issues that underlie international relations in this region. For many of these, such as the control of smuggling or the spread of infectious diseases, collaborative programs would enhance the national efforts.

Extending the logic, regional cooperation initiatives that could truly raise growth rates and living standards could provide incentives to put aside current behavior and the risks of regional conflict and could be a legitimate third option for the South China Sea. The experiences of existing regional cooperation initiatives show strongly that this is neither a simple nor a sure path—while some programs appear to encourage peaceful dispute resolution, others appear to have little real impact. Distilling these experiences and looking for programs that ride existing market trends suggest some principles for a possible initiative, one that would give life to this third option of peaceful development on the shores of the South China Sea.

Appendix A: Tourism at Risk from West Pacific Conflicts

Tourism represents part of the very rapidly growing service economy in East and Southeast Asia. Figure A.1 shows that, except for small oil and gas-exporting Brunei, tourism receipts total 2 % or larger of total exports of goods and services. For perspective, in 2014, the 10th largest commodity

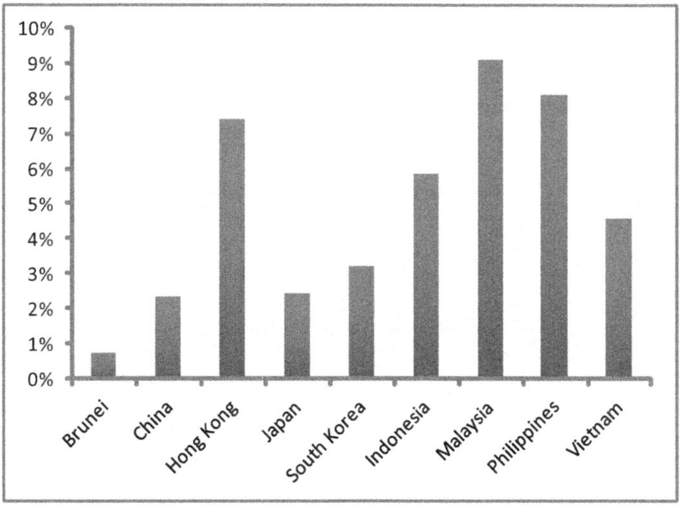

Fig. A.1 Tourism earnings/total exports (2013).
Notes: Data from World Bank web site: World Development Indicators. Brunei is 2012.

D.J. Green, *The Third Option for the South China Sea*,
DOI 10.1007/978-3-319-40274-1

export in the huge Chinese economy, iron or steel products, accounted for 2.6% of total exports (Workman 2016).

These service earnings can be strongly affected by the conflicts in the West Pacific Ocean. Historically, South Korean travelers to Japan represent the largest national contingent of visitors to that country. Figure A.2 shows the visitor arrivals in Japan from South Korea were down sharply in 2008 as a result of an uptick in the Dokdo/Takeshima dispute. This drop was comparable to that experienced after the 2011 earthquake and tsunami in Japan. Similarly, there is a clear fall-off in travelers from China to Japan in mid-2012 corresponding to the heightened political tension between the two countries in that year.

Annual data on inward-bound travelers to China and Hong Kong from Japan and South Korea, respectively, are less clear-cut. In addition to East China Sea disputes, travel from Japan in recent years was affected by the earthquake and tsunami in 2011, as well as cyclical income changes. The analogous picture of South Korean travel is also not pronounced. It is likely that future work on this subject needs to use higher frequency data.

Additional anecdotal information concerning how the West Pacific disputes may have influenced regional travel is given below:

- In Southeast Asia, in 2014 after the face-off in the Paracel islands over Chinese oil exploration activities, Chinese visitors to Viet Nam dropped precipitously (Thanh Nien News 2015). This decline

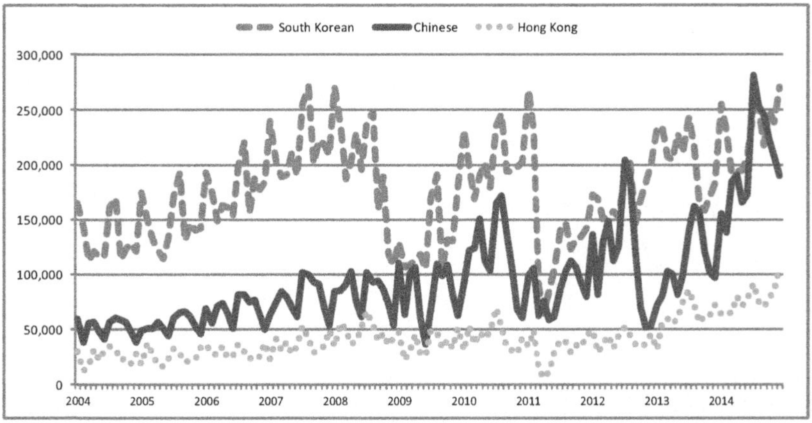

Fig. A.2 Visitor arrivals in Japan.
Notes: JTM, monthly data, Includes business travel.

continued into 2015. In 2015-Q1, Chinese travelers to Viet Nam were down 40.4% from the comparable period in 2014 (Viet Nam Tourism 2015).

- Dawson (2012) notes that mid–2012 commentaries expected a drop-off of outward-bound tourists from Japan to China as a result of heightened political tensions.
- In mid-2012, the Chinese–Philippines dispute at the Scarborough Shoal resulted in widely publicized cancelation of Chinese tourists traveling to the Philippines (Johanson 2012).

To-date, the impacts of the maritime conflicts have been relatively short-lived; however, this would not necessarily be the case were the conflicts to worsen.

The material in this section is provisional; there is considerable work that needs to be done to understand how the political tensions and the maritime disputes are affecting this important sector of the broader economies in the region.

Appendix B: Literature References for Policy Issues for the South China Sea

Table B.1 References to national policy issues for the South China Sea disputes

Issues	*Notes; Illustrative examples*
Provide strategic space for security forces	"… China appears to require greater control over the South China Sea in order to guarantee its security" Cronin (2013, p. 1). China has "the need for 'strategic depth' to protect China's coastal cities" Hayton (2014, p. 252). Particularly for China, with its expanding world economic and political power and presence, it is axiomatic that there would be pressure to expand its present defensive perimeter to what is sometimes referred to as First Island Chain—roughly from southern Japan, snaking around Taiwan to include most of the South China Sea. See, for instance, the discussion in Xiaokun (2013) and Cronin (2013, p. 1). Hong (2012) reviews the China–Philippines dispute from security aspects (in addition to economic interests). Zhong (2013) emphasizes the military or defense benefits prompting the disputed claims.
Enforce sovereignty, control of territory	Indonesia views the need to develop the capacity to defend its South China Sea claims as one aspect of the broader need to ensure maritime security Shekhar and Liow (2014). China "wants to assert what it sees as its historic rights, including sovereignty over all the geographical features and possibly even the entire maritime space" Storey (2012, p.53). The Chinese Government sees a successful effort to claim maritime rights providing the authority to restrict the transit of the naval vessels from other countries - The Economist (2014c).

(*continued*)

D.J. Green, *The Third Option for the South China Sea*,
DOI 10.1007/978-3-319-40274-1

Table B.1 (continued)

Issues	*Notes; Illustrative examples*
Projecting an image of strength	China's efforts in the South China Sea stem partly from "a desire for national prestige" Hayton (2014, p. 252).
Access to mineral, especially energy, resources	"Maritime rights involve…the right to exploit whatever resources are contained in the water column and seabed (especially petroleum..." Fravel (2012, p. 34). "Fanned by hoped-for discoveries of oil and gas, seabed territorial disputes fester" Frost (2008, p. 10).
Disaster relief	China's policy toward South China Sea includes the stated goal of "disaster prevention and mitigation" Stoa (2015). Chinese island building will help China support "disaster prevention and mitigation" Lubold and Entous (2015).
Protecting the environment, reduce water pollution	China's policy toward South China Sea includes the stated goal of "environmental protection" Stoa (2015). "Dense clouds of smoke haze were widely evident in satellite photo-images of the South China Sea during the last half of 1997" Rosenberg (SCS online).
Counterterrorism	While not common, "the high possible cost still makes maritime terrorism a substantial risk" Rosenberg (2009, p. 49).
Control illegal trade, trafficking in people and smuggling of controlled goods and substances; enforcing sanitary and phytosanitary rules	"The range of criminal activity around seaports is extensive, including the smuggling or illicit import of illegal drugs, contraband, stowaways and aliens, restricted or prohibited merchandise, and munitions" Rosenberg (2009, p. 48). "The agreement on the establishment of a hotline between the Philippine Coast Guard and the Vietnam Marine Police aims [at the]… prevention of smuggling, drug trafficking, illegal immigration and piracy" Bordadora (2011).
Combat piracy	"Piracy is an ancient, persistent, and elusive phenomenon in the South China Sea" Rosenberg (2009, p. 43). "[T]he area with the most pirate attacks in recent years has been the South China Sea" Schoenberger (2014).
Reduce illegal, unsustainable fishing	"[M]ost fish stocks in the western South China Sea are exploited or overexploited " Rogers (2012, p. 90). "Overfishing, illegal, unreported and unregulated (IUU) fishing" represent policy challenges in the South China Sea - Pejsova (2014, p. 1).
Ensuring freedom of transit and travel	China acts " to ensure that its sea lines of communication (SLOCs) are secure because these trade arteries are so vital" Storey (2012, p. 53). "'This is a very important issue, and has become the main concern of Japan, the United States and even right now the European Union,' said Dr. Yann-Huei Song" Xu (2014, p. 3).

(*continued*)

Table B.1 (continued)

Issues	*Notes; Illustrative examples*
Preventing the spread of infectious diseases	"World health officials are warning that diseases originating in Asia, such as avian flu, could give rise to the next global pandemic" Frost (2008, p. 6). "[W]e remain vulnerable to pandemic threats and garner the call to unity for continued vigilance in protecting our region from any disease threat" Pitsuwan (2009).

References

Asia Maritime Transparency Initiative (AMTI). 2015a. A fiery cross to bear. http://amti.csis.org/fiery-cross/ Accessed 4 Mar 2016.

———. 2015b. Island tracker. http://amti.csis.org/island-tracker/. Accessed 4 March 2016.

———. 2015c. Sandcastles of their own: Vietnamese expansion in the Spratly Islands. http://amti.csis.org/vietnam-island-building/. Accessed 4 Mar 2016.

Asian Development Bank (ADB). 2009a. Corridor chronicles: Profiles of cross border activities in the Greater Mekong subregion. http://adb.org/sites/default/files/pub/2009/corridor-chronicles-gms.pdf. Accessed 23 June 2015.

———. 2009b. REG: Greater Mekong subregion economic cooperation program: Joint ministerial statement: 15th ministerial conference, June 2009. http://www.adb.org/sites/default/files/page/42450/15th-summit-joint-ministerial-conference-statement-greater-mekong-subregion-gms.pdf. Accessed 23 June 2015.

———. 2010. Bridges across oceans: Initial impact assessment of the Philippines Nautical Highway System and lessons for Southeast Asia. [online] Publication, April 2010. http://www.adb.org/publications/bridges-across-oceans-initial-impact-assessment-philippines-nautical-highway-system-and. Accessed 14 Aug 2015.

———. 2011. Support for Pan-Beibu Gulf Economic Cooperation. Technical assistance report, Project number: 45002, Dec. 2011. http://www.adb.org/sites/default/files/project-document/60295/45002-001-reg-tar.pdf. Accessed 22 June 2015.

———. 2012. Project details: 46232–001: Greater Mekong subregion Phnom Penh Plan for Development Management Phase V. http://www.adb.org/projects/46232-001/main. Accessed 6 Apr 2016.

D.J. Green, *The Third Option for the South China Sea*,
DOI 10.1007/978-3-319-40274-1

———. 2013a. Asian development outlook 2013: Asia's energy challenge. http://www.adb.org/publications/asian-development-outlook-2013-asias-energy-challenge. Accessed 2 May 2015.

———. 2013b. Asian development outlook 2013 update. Governance and public service delivery. http://adb.org/sites/default/files/pub/2013/ado2013-update.pdf. Accessed 13 May 2015.

———. 2013c. ADB's first BIMP-EAGA project to bring clean energy to West Kalimantan. 28 Aug. 2013. News Release. http://www.adb.org/news/adbs-first-bimp-eaga-project-bring-clean-energy-west-kalimantan. Accessed 26 June 2015.

———. 2015. GMS e-Updates, September 2014–February 2015. Vol. 8, Issue No. 2. http://www.adb.org/sites/default/files/publication/154923/gms-e-updates-sep2014-feb2015.pdf. Accessed 18 June 2015.

Asian Development Bank Institute (ADBI). 2014. ASEAN 2030: Toward a borderless economic community. Tokyo, Japan. http://adb.org/sites/default/files/pub/2014/asean-2030-toward-borderless-economic-community.pdf. Accessed 24 Aug 2015.

Associated Press (AP). 2015. Indonesia sinks 41 foreign boats to warn against poaching. Yahoo News, May 20, 2015. http://news.yahoo.com/indonesia-sinks-41-foreign-boats-warn-against-poaching-144052862.html. Accessed 25 Mar 2016.

Association of Southeast Asian Nations (ASEAN DOC). 2002. Declaration on the conduct of parties in the South China Sea. http://www.asean.org/?static_post=declaration-on-the-conduct-of-parties-in-the-south-china-sea-2. Accessed 23 Mar 2016.

Association of Southeast Asian Nations (ASEAN, AEC). ASEAN Economic Community. http://www.asean.org/communities/asean-economic-community. Accessed 3 Apr 2016.

Association of Southeast Asian Nations (ASEAN, APSC). ASEAN Political—Security Community. http://www.asean.org/communities/asean-political-security-community. Accessed 7 Apr 2015.

ASEAN UP. 2014. Benefits of the ASEAN Economic Community—AEC. ASEAN UP. http://aseanup.com/benefits-asean-economic-community-aec/. Accessed 7 Apr 2015.

Asia Regional Integration Center website. http://aric.adb.org/

Bader, Jeffrey, Kenneth Lieberthal, and Michael McDevitt. 2014. Keeping the South China Sea in perspective. Brookings, August 2014. http://www.brookings.edu/research/papers/2014/08/south-china-sea-perspective-bader-lieberthal-mcdevitt. Accessed 20 May 2015.

Baker, Rodger. 2016. Fish: The overlooked destabilizer in the South China Sea, 12 Feb. 2106. Stratfor. https://www.stratfor.com/analysis/fish-overlooked-destabilizer-south-china-sea. Accessed 21 Mar 2016.

Batam Indonesia Free Zone Authority (BIFZA). Batam's economic indicator period 2010–2014. Bpbatam.go.id. http://www.bpbatam.go.id/eng/Industry_economy/indicator.jsp. Accessed 18 June 2015.

Bernard, Leonardo. 2013. Prospect for joint development in the South China Sea. [online] "Managing tensions in the South China Sea" conference held by CSIS on June 5–6, 2013. http://csis.org/files/attachments/130606_Bernard_ConferencePaper.pdf. Accessed 22 Nov 2014.

Bhattacharya, Abheek. 2014. Japan car makers Skid in China, 3 Nov. 2013. *The Wall Street Journal.* http://online.wsj.com/articles/japan-carmakers-skid-in-china-1415009128. Accessed 6 Nov 2014.

Bordadora, Norman. 2011. Philippines, Vietnam forge naval agreement on Spratlys. Philippine Daily Inquirer, 27 Oct 2011. http://globalnation.inquirer.net/16517/philippines-vietnam-forge-naval-agreement-on-spratlys. Accessed 21 June 2015.

Bradsher, Keith. 2010. Amid tension, China blocks vital exports to Japan. The New York Times. 22 Sept 2010. http://www.nytimes.com/2010/09/23/business/global/23rare.html?pagewanted=all&_r=0. Accessed 7 Dec 2014.

———. 2014. Looking beyond China, some companies shift personnel. The New York Times, 9 Sept 2014 http://www.nytimes.com/2014/09/10/business/international/looking-beyond-china-some-companies-shift-personnel.html?_r=0. Accessed 10 Feb 2015.

British Broadcasting Corporation (BBC). 2012. Asean nations fail to reach agreement on South China Sea, 13 July 2012. http://www.bbc.com/news/world-asia-18825148. Accessed 10 Apr 2015.

———. 2013. Q&A: Thailand-Cambodia temple dispute. BBC News, 7 Nov. 2013. http://www.bbc.com/news/world-asia-pacific-12378001. Accessed 23 June 2015.

———. 2014. How uninhabited islands soured China-Japan ties. BBC News, 10 Nov 2014. http://www.bbc.com/news/world-asia-pacific-11341139. Accessed 16 Dec 2014.

———. 2015a. Q&A: South China Sea dispute. BBC News, 27 Oct 2015. http://www.bbc.com/news/world-asia-pacific-13748349. Accessed 22 July 2015.

———. 2015b. "Australia conducting 'freedom of navigation' flights In South China Sea—BBC News, 15 Dec 2015. http://www.bbc.com/news/world-australia-35099445. Accessed 19 Mar 2016.

Brunei Darussalam-Indonesia-Malaysia-Philippines–East ASEAN Growth Area (BIMP). 2012. Implementation blueprint 2012–2016, March 2012. http://www.adb.org/sites/default/files/page/34232/bimp-eaga-implementation-blueprint-2012-2016.pdf. Accessed 6 Apr 2016.

Bunnell, Tim, Carl Grundy-Warr, James D. Sidaway, and Matthew Sparke. 2012. Geographies of power in the Indonesia-Malaysia-Singapore growth triangle. In *International handbook of globalization and world cities*, ed. B. Derudder, M. Hoyler, P.J. Taylor, and F. Witlox, 465–475. Cheltenham/Northampton:

Edward Elgar. http://faculty.washington.edu/sparke/geopower.pdf. Accessed 18 June 2015.

Buszynski, Leszek. 2011. The South China Sea: Avenues towards a resolution of the issue. Southchinaseastudies.org, 24 March 2011. http://southchinaseastudies.org/en/conferences-and-seminars-/515-the-south-china-sea-avenue-towards-a-resolution-of-the-issue. Accessed 21 May 2015.

Calleja, Niña P. 2016. Chinese vessels have left Quirino Atoll—DFA. Philippine Daily Inquirer, 2 Mar 2016. http://globalnation.inquirer.net/137219/chinese-vessels-have-left-quirino-atoll-dfa-dnd. Accessed 6 Mar 2016.

Center for a New American Security (CNAS). 2014. Timeline: 1955-Present. Cnas.org. http://www.cnas.org/flashpoints/timeline. Accessed 3 Jan 2015.

Center for Strategic and International Studies (CSIS). 2015. 21 Jul. 2015, Washington, DC. http://csis.org/event/fifth-annual-csis-south-china-sea-conference. Accessed 16 Feb 2016.

———. 2016. Asia-Pacific rebalance 2025 capabilities, presence, and partnerships: An independent review of U.S. defense strategy in the Asia-Pacific. http://csis.org/files/publication/160119_Green_AsiaPacificRebalance2025_Web_0.pdf. Accessed 11 Mar 2016.

Chang, Felix K. 2014. Comparative Southeast Asian military modernization—1. The Asan Forum: Topics of the month, 01 Oct 2014. http://www.theasanforum.org/comparative-southeast-asian-military-modernization-1/. Accessed 18 May 2015.

Chia, Siow Yue. 1997. Regionalism and subregionalism in ASEAN: The free trade area and growth triangle models. In *Regionalism versus multilateral trade arrangements*, NBER-EASE Volume 6, ed. Takatoshi Ito and Anne O. Krueger, 275–312. University of Chicago Press. http://www.nber.org/chapters/c8603. Accessed 6 Apr 2016.

(China) National Bureau of Statistics of China. 2014. China statistical yearbook-2014. http://www.stats.gov.cn/tjsj/ndsj/2014/indexeh.htm. Accessed 20 Mar 2016.

Choi, Byung-il, and Changyong Rhee. 2014. Future of factory Asia. Asian Development Bank and Korea Economic Research Institute. http://www.adb.org/sites/default/files/publication/42477/future-factory-asia.pdf. Accessed 31 Mar 2016.

Council on Foreign Relations (CFR). 2016. China's maritime disputes. A CFR InfoGuide Presentation. http://www.cfr.org/asia-and-pacific/chinas-maritime-disputes/p31345#!/p31345. Accessed 6 Mar 2016.

Cronin, Patrick M. (ed). 2012 Cooperation from strength: The United States, China and the South China Sea. Washington, DC: Center for a New American Security. http://www.cnas.org/sites/default/files/publications-pdf/CNAS. Accessed 16 May 2015.

Cronin, Patrick M. 2013. The strategic significance of the South China Sea. For "Managing tensions in the South China Sea" conference held by the Center for

Strategic & International Studies (CSIS) on 5–6 June 2013. http://csis.org/files/attachments/130606_Cronin. Accessed 23 Nov 2014.
———. 2014. The challenge of responding to maritime coercion. [online]] Washington, DC: Center for a New American Security, September 2014. http://www.cnas.org/Challenge-Responding-to-Maritime-Coercion#.Vu8M0xIrJDU. Accessed 20 Mar 2016.
Cronin, Patrick M., and Robert D. Kaplan. (2012. Cooperation from strength: U.S. strategy and the South China Sea. In *Cooperation from strength: The United States, China and the South China Sea*, ed. Cronin, 3–30. Washington, DC: Center for a New American Security. http://www.cnas.org/sites/default/files/publications-pdf/CNAS. Accessed 16 May 2015.
Currie, Kelley. 2010. Why is China picking fights with Indonesia? The Weekly Standard. http://www.weeklystandard.com/blogs/why-china-picking-fights-indonesia. Accessed 4 Jan 2015.
Das, Sanchita Basu. 2012. ASEAN: A united front To tackle The South China Sea issue. East Asia Forum. http://www.eastasiaforum.org/2012/05/13/asean-a-united-front-to-tackle-the-south-china-sea-issue/. Accessed 13 May 2012.
Das, Sanchita Basu, Jayant Menon, Rodolfo C. Severino, and Omkar Lal Shrestha (eds). 2013. The ASEAN economic community: A work in progress, Asian Development Bank and Institute of Southeast Asian Studies. http://www.adb.org/sites/default/files/publication/31147/aec-work-progress.pdf. Accessed 1 Jan 2016.
Das, Satyajit. 2014. China's debt vulnerability. EconoMonitor, 9 Apr 2014. http://www.economonitor.com/blog/2014/04/chinas-debt-vulnerability/. Accessed 16 Dec 2014.
Dawson, Chester. 2012. Japan's spat with China takes big toll on tourism. The Wall Street Journal, 27 Nov 2012. http://www.wsj.com/articles/SB1000142412788732407350457810814180951 2474. Accessed 10 May 2015.
De Castro, Renato Cruz. 2013. China's realpolitik approach in the South China Sea dispute: The case of the 2012 Scarborough Shoal stand-off. For Managing Tensions in the South China Sea conference held by the Center for Strategic & International Studies (CSIS) on 5–6 June 2013. http://csis.org/files/attachments/130606_DeCastro_ConferencePaper.pdf. Accessed 23 Nov 2014.
DevonshireEllis, Chris. 2014. China-ASEAN wage comparisons and the 70 percent production capacity benchmark. China Briefing, Dezan Shira & Associates, 3Jun2014.http://www.china-briefing.com/news/2014/06/03/china-asean-wage-comparisons-70-production-capacity-benchmark.html. Accessed 1 Aug 2015.
Djalal, Hasjim. 2011. Rethinking the zone of peace, freedom and neutrality (ZOPFAN) in the Post-Cold War era. 25th Asia-Pacific Roundtable, 31 May 2011. http://www.isis.org.my/files/25APR/paper_ps_5_hasjim_djalal.pdf. Accessed 17 Jun 2015.

Domínguez, Gabriel. 2014. Japan shifting investments from China to ASEAN. Deutsche Welle, 25 Jul 2014. http://www.dw.de/japan-shifting-investments-from-china-to-asean/a-17809068. Accessed 2 Oct 2014.

Dosch, Jörn, and Oliver Hensengerth. 2005. Sub-regional cooperation in Southeast Asia: The Mekong Basin. *European Journal of East Asian Studies* 4(2): 263–286. http://www.wiwi.uni-rostock.de/fileadmin/Institute/IPV/Lehrstuehle/Internationale_Politik/J%C3%B6rn_Dosch/Dosch_Hensengerth.pdf. Accessed 23 Jun 2015.

The Economist. 2013. China's outward investment: The second wave, 26 Oct 2013. http://www.economist.com/news/business/21588369-what-make-chinese-firms-latest-foreign-purchases-second-wave. Accessed 17 Feb 2015.

———. 2014a. Not the usual drill. 10 May 2014. http://www.economist.com/news/asia/21601879-tensions-mount-dangerously-contested-waters-not-usual-drill. Accessed 17 Nov 2014.

———. 2014b. Rigged: Vietnam and the South China Sea. 14 Jun 2014. http://www.economist.com/news/asia/21604203-comradely-relations-go-bad-worse-rigged. Accessed 11 Jan 2015.

———. 2014c. Maritime power, your rules or mine? 15 Nov 2014. http://www.economist.com/news/special-report/21631792-trade-depends-order-sea-keeping-it-far-straightforward-your-rules-or. Accessed 6 May 2015.

———. 2015. Small reefs, big problems. 25 Jul 2015. http://www.economist.com/news/asia/21659771-asian-coastguards-are-front-line-struggle-check-china-small-reefs-big-problems. Accessed 1 Aug 2015.

Encyclopedia Britannica. 2013. Paracel Islands | islands, South China Sea. Encyclopedia Britannica, 21 Jul 2013. http://www.britannica.com/EBchecked/topic/442423/Paracel-Islands. Accessed 3 Jan 2015.

Erickson, Andrew S., and Austin Strange. 2014. Pandora's Sandbox: China's Island-building strategy in the South China Sea. *Foreign Affairs* 13 Jul 2014. https://www.foreignaffairs.com/articles/china/2014-07-13/pandoras-sandbox. Accessed 6 Mar 2016.

Food and Agriculture Organization (FAO). 2012. FAO yearbook. Fishery and aquaculture statistics. FAO Fisheries and Aquaculture Department. http://www.fao.org/3/a-i3740t/index.html. Accessed 5 Apr 2016.

Francisco, Rosemarie. 2014. Geopolitical tensions could hurt Asia growth, investments: IMF deputy MD. Reuters 21 May 2014. http://www.reuters.com/article/2014/05/21/us-imf-asia-risks-idUSBREA4K0B820140521. Accessed 6 May 2015.

Fravel, M. Taylor. 2012. Maritime security in the South China Sea and the competition over maritime rights. In *Cooperation from strength the United States, China and the South China Sea*, ed. P. Cronin. Washington, DC: Center for a New American Security, 31–50. http://www.cnas.org/sites/default/files/publications-pdf/CNAS. Accessed 16 May 2015.

Friedman, George. 2014. Viewing Russia from the inside. Stratfor. http://www.stratfor.com/weekly/viewing-russia-inside#axzz3PDlLF1uL. Accessed 19 Jan 2015.

———. 2016. The significance of US, Indian and Japanese naval exercises. Geopolitical Futures, 4 Mar 2016. https://geopoliticalfutures.com/the-significance-of-us-indian-and-japanese-naval-exercises/. Accessed 20 Mar 2016.

Frost, Ellen L. 2008. *Asia's new regionalism*. Boulder: Lynne Rienner.

GMA News Online. 2011. Narrower 'right-hand' drive ban to run over truck, bus imports? Money. http://www.gmanetwork.com/news/story/210733/money/narrower-right-hand-drive-ban-to-run-over-truck-bus-imports. Accessed 22 Feb 2016.

Green, David Jay. 2007. Bridging the ASEAN developmental divide: Challenges and prospects, a regional overview. *ASEAN Economic Bulletin* April(24/#1): 15–34.

———. 2008a. Alphabet soup and the ASEAN neighborhoods. *The Jakarta Post*, 4 Jan 2008. http://www.adb.org/news/op-ed/alphabet-soup-and-asean-neighborhoods. Accessed 15 June 2015.

———. 2008b. The role of ASEAN economic community as a commitment to policy certainty. *ASEAN Economic Bulletin* August(25/#2): 209–227.

———. 2010. Southeast Asia's policy response to the global economic crisis. *ASEAN Economic Bulletin* April(27/#1): 5–26.

———. 2013. Fighting over the West Philippine Sea is so 17th century. *Business Mirror*, Manila, 30 June 2013.

Haddick, Robert. 2012. Salami slicing in the South China Sea. *Foreign Policy*, 3 Aug. 2012. http://foreignpolicy.com/2012/08/03/salami-slicing-in-the-south-china-sea/. Accessed 23 Oct 2015.

Hayashi, Yuka, and Mayumi Negishi. 2013. Japan's companies Shun China for Southeast Asia. *The Wall Street Journal*, 12 Sept 2013. http://www.wsj.com/articles/SB10001424127887324549004579070820138376020. Accessed 10 Feb 2015.

Hayton, Bill. 2014. *The South China Sea: The struggle for power in Asia*. New Haven: Yale University Press.

Holmes, James R. 2012. Rough waters for coalition building. In *Cooperation from strength the United States, China and the South China Sea*, ed. P. Cronin, 99–115. Washington, DC: Center for a New American Security. http://www.cnas.org/sites/default/files/publications-pdf/CNAS. Accessed 16 May 2015.

Hong, Zhao. 2012. Sino-Philippines relations: Moving beyond South China Sea dispute? *The Journal of East Asian Affairs* 26(2(Fall/Winter)): 57–76.

———. 2013. China's FDI into Southeast Asia. ISEAS perspective, (Institute of Southeast Asian Studies, Singapore). #08, 31 Jan 2013.

Hookway, James. 2013. What's in a name? In the South China Sea, it seems, quite a lot. *The Wall Street Journal*, 9 Oct 2013. http://blogs.wsj.com/indonesiarealtime/2013/10/09/whats-in-a-name-in-the-south-china-sea-it-seems-quite-a-lot/?mg=blogs-wsj&url=http%3A%2F%2Fblogs.wsj.com%2Fsearealtime%2F2013%2F10%2F09%2Fwhats-in-a-name-in-the-south-china-sea-it-seems-quite-a-lot%2F. Accessed 16 Dec 2014.

Hosokawa, Daisuke. 2009. Pan-Beibu Gulf Economic Cooperation—China's new initiative in cooperation with ASEAN. *Osaka Keidai Ronshu* 60(2 July): 67–78. www.osaka-ue.ac.jp/file/general/4626. Accessed 11 Aug 2015.

Huyen, Hoang Thu, and My Nguyen. 2013. Development zones in Vietnam. Vietnam Briefing News, 4 Jun 2013. http://www.vietnam-briefing.com/news/development-zones-vietnam.html/. Accessed 7 Apr 2016.

Indonesia-Malaysia-Thailand--Growth Triangle (IMT-GT). 2012. Implementation blueprint 2012–2016. http://imtgt.org/Documents/IMT-GT_Implementation_Blueprint_2012-2016.pdf. Accessed 27 Jun 2015.

International Crisis Group (ICG). 2012. Stirring up the South China Sea (II): Regional responses, report N 229, 24 July 2012. http://www.crisisgroup.org/~/media/Files/asia/north-east-asia/229-stirring-up-the-south-china-sea-ii-regional-responses. Accessed 22 May 2015.

International Monetary Fund. 2014a. Republic of Korea: 2013 Article IV Consultation—Staff Report, Press release and statement by the Executive Director for the Republic of Korea. IMF country report no. 14/101. April 2014. https://www.imf.org/external/pubs/ft/scr/2014/cr14101.pdf. Accessed 6 May 2015.

———. 2014b. Brunei Darussalam: Statistical appendix. IMF country report no. 14/191. July 2014. https://www.imf.org/external/pubs/ft/scr/2014/cr14191.pdf. Accessed 6 May 2015.

———. 2014c. People's Republic of China: 2014 Article IV Consultation—Staff Report; press release; and statement by the Executive Director for the People's Republic of China: IMF country report no. 14/235. July 2014. https://www.imf.org/external/pubs/ft/scr/2014/cr14235.pdf. Accessed 6 May 2015.

———. 2014d. Japan: 2014 Article IV Consultation—Staff Report; and press release. IMF country report no. 14/236. July 2014. https://www.imf.org/external/pubs/ft/scr/2014/cr14236.pdf. Accessed 6 May 2015.

———. 2014e. Philippines: 2014 Article IV Consultation—Staff Report. IMF country report; and press release. No. 14/245. August 2014. https://www.imf.org/external/pubs/ft/scr/2014/cr14245.pdf. Accessed 6 May 2015.

———. 2014f. Vietnam: 2014 Article IV Consultation—Staff Report. IMF country report—Staff report; press release; and statement by the Executive Director for Vietnam. No. 14/311. October 2014. https://www.imf.org/external/pubs/ft/scr/2014/cr14311.pdf. Accessed 6 May 2015.

———. 2015a. Malaysia 2014 Article IV Consultation—Staff report; press release; and statement by the Executive Director for Malaysia. IMF country report no. 15/58. March 2015. https://www.imf.org/external/pubs/ft/scr/2015/cr1558.pdf. Accessed 6 May 2015.

———. 2015b. Indonesia: 2014 Article IV Consultation—Staff report; press release; and statement by the Executive Director for Indonesia. IMF country report no. 15/74. March 2015. https://www.imf.org/external/pubs/ft/scr/2015/cr1574.pdf. Accessed 6 May 2015.

———. 2015c. IMF Executive Board Concludes 2015 Article IV Consultation with Brunei Darussalam. Press release no. 15/259. June 4, 2015. https://www.imf.org/external/np/sec/pr/2015/pr15259.htm. Accessed 4 Apr 2016.

Japan Tourism Marketing Co. (JTM). Statistical data web site. http://www.tourism.jp/en/statistics/. Accessed 10 May 2015.

Jennings, Ralph. 2013. South China Sea: Taiwan Snaps back as big fish circle. Forbes, 3 Sep 2013. http://www.forbes.com/sites/ralphjennings/2013/09/03/taiwan-snaps-back-as-bigger-fish-circle/#1b96f5452753. Accessed 5 Mar 2016.

Johanson, Mark. 2012. China-Philippines South China Sea dispute hurting Philippines? Tourism industry. International Business Times, 14 May 2012. http://www.ibtimes.com/china-philippines-south-china-sea-dispute-hurting-philippines-tourism-industry-698300. Accessed 10 May 2015.

Khemakorn, Pakjuta. 2006. Sustainable management of pelagic fisheries in the South China Sea Region, November 2006. http://www.un.org/depts/los/nippon/unnff_programme_home/fellows_pages/fellows_papers/khemakorn_0607_thailand.pdf. Accessed 17 May 2015.

Kolstad, Ivar, and Arne Wiig. 2009. What determines Chinese outward FDI? Chr. Michelsen Institute. CMI working paper WP 2009: 3. Bergen, Norway. http://www.cmi.no/publications/publication/?3332=what-determines-chinese-outward-fdi. Accessed 16 Feb 2015.

Kyozuka, Tamaki. 2014. China's influence campaign: Beijing pushes 'go south' policy at Greater Mekong summit. Nikkei Asian Review. Asia.nikkei.com, 20 December 2014. http://asia.nikkei.com/print/article/66589. Accessed 5 Jul. 2015.

Lee, Soon Ho. 2013. Japan's dilemma: maritime disputes in East Asia. East Asia Forum, 2 March 2013. http://www.eastasiaforum.org/2013/03/02/japans-dilemma-maritime-disputes-in-east-asia/. Accessed 29 Jan 2015.

Lubold, Gordon, and Adam Entous. 2015. U.S. says Beijing is building up South China Sea Islands. *The Wall Street Journal*, May 9, 2015. http://www.wsj.com/articles/u-s-says-beijing-building-up-south-china-sea-islands-1431109387. Accessed 9 Aug 2015.

Lubold, Gordon, and Jeremy Page. 2015. U.S. navy ship sails near islands claimed by China. *The Wall Street Journal*, Oct. 26, 2015. http://www.wsj.com/articles/u-s-navy-ship-sails-near-islands-claimed-by-china-1445908192?alg=y. Accessed 30 Oct 2015.

McLaughlin, Kathleen E. 2011. South China Sea: China vs. Vietnam. *GlobalPost*, 14 Jun 2011. http://www.globalpost.com/dispatch/news/regions/asia-pacific/china/110614/south-china-sea-china-vs-vietnam. Accessed 20 Jan 2015.

Ministry of Foreign Affairs. People's Republic of China (MOFA, China). 2016. Set aside dispute and pursue joint development. http://www.fmprc.gov.cn/mfa_eng/ziliao_665539/3602_665543/3604_665547/t18023.shtml. Accessed 7 Apr 2016.

Mogato, Manuel. 2015. From rundown outpost, Philippines watches China Island take shape in disputed sea. Reuters. 11 May 2015. http://www.reuters.com/article/us-southchinasea-philippines-outpost-idUSKBN0NW18F20150511. Accessed 5 Mar 2016.

Mullen, Jethro. 2013. Malaysia launches attack on Filipino intruders in Borneo. CNN, 5 Mar 2013. http://www.cnn.com/2013/03/04/world/asia/malaysia-philippines-standoff/. Accessed 3 Jul 2015.

Nguyen Thai Hoc Foundation. 2014. Change the name "South China Sea" to "Southeast Asia Sea", Change.org. https://www.change.org/p/change-the-name-south-china-sea-to-southeast-asia-sea. Accessed 16 Dec 2014.

New York Times. 2012. Territorial claims in South China Sea. Nytimes.com, 31 May 2012. http://www.nytimes.com/interactive/2012/05/31/world/asia/Territorial-Claims-in-South-China-Sea.html?_r=0. Accessed 3 Jan 2015.

Nor-Afidah. 2002. Growth triangle | Infopedia. National Library Board Singapore. http://eresources.nlb.gov.sg/infopedia/articles/SIP_58_2005-01-06.html?utm_expid=85360850-6.qNOOYF40RhKK6gXsQEaAJA.0&utm_referrer=https%3A%2F%2Fwww.google.com%2F. Accessed 18 Jun 2015.

O'Shea, Paul. 2013. Territorial disputes in Northeast Asia: A primer. Italian Institute for International Political Studies Analysis (ISPI) No. 182, June 2013. http://www.ispionline.it/sites/default/files/pubblicazioni/analysis_182_2013.pdf. Accessed 8 Dec 2014.

Palmerston, Henry John Temple. (1848). Remarks in the house of commons, the quotations page. http://www.quotationspage.com/quote/41290.html. Accessed 17 Feb 2016.

Pejsova, Eva. 2014. The South China Sea's commons: Behind and beyond sovereignty disputes. European Union Institute for Security Studies, June 2014. http://www.iss.europa.eu/uploads/media/Alert_30_South_China_Sea.pdf. Accessed 17 May 2015.

Pitsuwan, H.E. Dr Surin. 2009. Speech by Secretary-General of ASEAN at the ASEAN Plus Three Health Ministers Special Meeting on Influenza A(H1N1), 8 May 2009. http://www.asean.org/?static_post=speech-by-he-dr-surin-pitsuwan-secretary-general-of-asean-at-the-asean-plus-three-health-ministers-special-meeting-on-influenza-ah1n1. Accessed 7 Aug 2015.

Quintos, Mary Fides A. 2015. Artificial islands in the South China Sea and their impact on regional (in)security. FSI Insights, Center for International Relations and Strategic Studies (CIRSS) of the Foreign Service Institute (FSI), Mar. 2015 (II/#2). http://www.fsi.gov.ph/wp-content/uploads/2015/03/2015-0305-Vol-2-No-2-FSI-Insights-Artificial-Islands-in-the-South-China-Sea-Qui ntos.pdf. Accessed 5 Mar 2016.

Ramos, Fidel. 1994. Cross-border growth triangles promote prosperity in East Asia. International Herald Tribune. 30 Nov 1994. Nytimes.Com. http://www.nytimes.com/1994/11/30/opinion/30iht-edramos.html. Accessed 16 Aug 2015.

Roach, Stephen. 2013. China's last soft landing? Project Syndicate, 29 Jan 2013. http://www.project-syndicate.org/commentary/chinese-economic-resilience-is-weakening-by-stephen-s--roach. Accessed 16 Dec 2014.

Roberts, Dexter. 2012. Anti-Japanese protests flare in China over disputed islands. Bloomberg Businessweek, 17 Sept 2012. http://www.businessweek.com/articles/2012-09-17/anti-japanese-protests-flare-in-china-over-disputed-islands. Accessed 7 Dec 2014.

Roehrig, Terence. 2012. South Korea–China maritime disputes: Toward a solution. 27 Nov 2012. East Asia Forum. http://www.eastasiaforum.org/2012/11/27/south-korea-china-maritime-disputes-toward-a-solution/. Accessed 1 Dec 2014.

Rogers, Will. 2012. The role of natural resources in the South China Sea. In *Cooperation from strength the United States, China and the South China Sea*, ed. P. Cronin, 83–97. Washington, DC: Center for a New American Security. http://www.cnas.org/sites/default/files/publications-pdf/CNAS. Accessed 16 May 2015.

Rosenberg, David. 2009. The political economy of piracy in the South China Sea. *Naval War College Review*, Summer 2009, *62*(3): 43–58. Usnwc.edu. https://www.usnwc.edu/getattachment/cd4a4cfa-eaa0-4040-9d51-a4b90b7a83d5/The-Political-Economy-of-Piracy-in-the-South-China. Accessed 18 May 2015.

Rosenberg, David. Rosenberg SCS online. The South China Sea. http://www.southchinasea.org/. Accessed 17 May 2015.

Schoenberger, Sarah. 2014. Piracy in the South China Sea: Petty theft in Indonesia, kidnapped ships in Malaysia. Center for International Maritime Security, 6 Sept 2014. http://cimsec.org/piracy-south-china-sea-petty-theft-indonesia-kidnapped-ships-malaysia/12899. Accessed 18 May 2015.

Schotter, Andrew. 1990. Oskar Morgenstern's contribution to the development of the theory of games. http://cess.nyu.edu/schotter/wp-content/uploads/2010/02/%E2%80%9COskar-Morgenstern%E2%80%99s-Contribution-to-the-Development-of-the-Theory-of-Games%E2%80%9D.pdf. Accessed 16 Feb 2016.

Schwarz, Adam, and Roland Villinger. 2004. Integrating Southeast Asia's economies. *McKinsey Quarterly* (# 1/2004): 36–46. http://unpan1.un.org/intradoc/groups/public/documents/apcity/unpan049156.pdf. Accessed 11 Apr 2015.

Shear, David. Statement of David Shear Assistant Secretary of Defense for Asian & Pacific Security Affairs before the Senate Committee on Foreign Relations, May 13, 2015. U.S. Senate Committee on Foreign Relations. http://www.foreign.senate.gov/imo/media/doc/051315_Shear_Testimony.pdf. Accessed 6 Mar 2016.

Shekhar, Vibhanshu, and Joseph Chinyong Liow. 2014. Indonesia as a maritime power: Jokowi's vision, strategies, and obstacles ahead. The Brookings Institution. Nov 2014. http://www.brookings.edu/research/articles/2014/11/indonesia-maritime-liow-shekhar. Accessed 16 Dec 2014.

Shor, Mikhael. 2006. Game of chicken. Dictionary of Game Theory Terms. http://www.gametheory.net/dictionary/games/GameofChicken.html. Accessed 20 Feb 2016.

Shu-Ling, Ko. 2013. Details of new Japan-Taiwan fisheries pact are explained. The Japan Times, 23 Apr 2013. http://www.japantimes.co.jp/news/2013/04/23/national/details-of-new-japan-taiwan-fisheries-pact-are-explained/#.VVwhCGTBzGc. Accessed 20 May 2015.

Somanader, Tanya. 2015. President Obama: "Writing the rules for 21st century trade". The White House. https://www.whitehouse.gov/blog/2015/02/18/president-obama-writing-rules-21st-century-trade. Accessed 20 Apr 2015.

Spaniel, William. 2011. Game theory 101: Chicken. YouTube video. https://www.youtube.com/watch?v=sww-Zsl0IRY. Accessed16 Feb 2016.

Spegele, Brian, and Vu Trong Khanh. 2014. China, Vietnam try to repair ties after South China Sea dispute. *The Wall Street Journal.* 28 Aug 2014. http://www.wsj.com/articles/china-vietnam-try-to-repair-ties-after-oil. Accessed 11 Jan 2015.

Srivastava, Pradeep, and Utsav Kumar. 2012 GMS trade—trends and patterns. Chapter 2 (pp. 11–37). In *Trade and trade facilitation in the Greater Mekong Subregion*, ed. Srivastava, Pradeep and Utsav Kumar. Mandaluyong City: Asian Development Bank. http://www.adb.org/sites/default/files/publication/29977/trade-and-trade-facilitation-gms.pdf. Accessed 17 June 2015.

Stoa, Ryan. 2015. Environmental peacebuilding in the South China Sea. 1 & 9 May 2015 [Blog] http://ryanstoa.com/blog?category=South+China+Sea. Accessed 25 Mar 2016.

Storey, Ian. 2012. China's bilateral and multilateral diplomacy in the South China Sea. In *Cooperation from strength the United States, China and the South China Sea*, ed. P. Cronin, 51–66. Washington DC: Center for a New American Security. http://www.cnas.org/sites/default/files/publications-pdf/CNAS. Accessed 16 May 2015.

Stout, David. 2014. The last time China got into a fight with Vietnam, it was a disaster, 15 May 2014. TIME. http://time.com/100417/china-vietnam-sino-vietnamese-war-south-china-sea/. Accessed 17 Apr 2015.

Stratfor Global Intelligence. Stratfor, 2016. Japan wades further into the South China Sea dispute, 12 Jan 2016. Geopolitical Diary. https://www.stratfor.com/geopolitical-diary/japan-wades-further-south-china-sea-dispute. Accessed 21 Mar 2016.

(Editorial Team) Supplychains Magazine. 2012. Diaoyu Island dispute: Japanese supply chain hit hard, 10 Oct 2012. Supplychains Magazine.

http://supplychains.com/diaoyu-island-dispute-japanese-supply-chain-hit-hard/. Accessed 8 Dec 2014.

Sussangkarn, Chalongphob. 2012. Toward a functional Chiang Mai initiative. East Asia Forum, 15 May 2012. http://www.eastasiaforum.org/2012/05/15/toward-a-functional-chiang-mai-initiative/. Accessed 15 May 2015.

Taylor, Rob. 2016. Australia takes steps to counter China's rising military power. *The Wall Street Journal*, 24 Feb 2016. http://www.wsj.com/articles/australia-takes-steps-to-counter-chinas-island-building-1456366660. Accessed 19 Mar 2016.

Thanh Nien News. 2015. Vietnam tourism numbers plummet as Chinese, Russians keep away. 1 Apr 2015 Travel. http://www.thanhniennews.com/travel/vietnam-tourism-numbers-plummet-as-chinese-russians-keep-away-40558.html. Accessed 10 May 2015.

Thayer, Carlyle A. 2012. ASEAN'S Code of Conduct in the South China Sea: A litmus test for community-building? *The Asia-Pacific Journal* 10(34), no. 4, 20 Aug 2012. http://www.japanfocus.org/-Carlyle_A_-Thayer/3813. Accessed 8 Apr 2015.

———. 2015. South China Sea tensions: China, the claimant states, ASEAN and the major powers. In *Power, law, and maritime order in the South China Sea*, ed. T.T. Thuy and L.T. Trang, 3–35. Lanham: Lexington Books.

Thorbecke, Willem, and Nimesh Salike. 2011. Understanding foreign direct investment in East Asia. ADBI Working Paper 290, June 2011. Tokyo: Asian Development Bank Institute. http://www.adbi.org/working-paper/2011/06/23/4616.foreign.direct.investment.east.asia/. Accessed 1 Apr 2016.

Thuy, Tran Truong. 2015. Construction in the South China Sea: A comparative view. Asia Maritime Transparency Initiative. http://amti.csis.org/construction-in-the-south-china-sea-a-comparative-view/. Accessed 5 Mar 2016.

Thuy, Tran Truong, and Le Thuy Trang (eds.). 2015. *Power, law, and maritime order in the South China Sea*. Lanham: Lexington Books.

Tiezzi, Shannon. 2014. China discovers gas field in the South China Sea. The Diplomat, 16 Sept 2014. http://thediplomat.com/2014/09/china-discovers-gas. Accessed 16 May 2015.

Torode, Greg. 2013. Spratly Islands dispute defines China-Vietnam relations 25 years after naval clash. South China Morning Post, 23 Dec 2014. http://www.scmp.com/news/asia/article/1192472/spratly-islands-dispute-defines-china-vietnam-relations-25-years-after. Accessed 2 Jan 2015.

Trading Economies website. http://www.tradingeconomics.com

Tran, Du. 2009. The South China Sea dispute: An unbalanced bargaining game. The Southeast Asian Sea Research Foundation. http://en.seasfoundation.org/index.php?option=com_content&view=article&id=48:the-south-china-sea-dispute-an-unbalanced-bargaining-game&catid=1:latest-news&Itemid=50. Accessed 16 Nov 2014.

Twain. 1904. From *Mark Twain's own autobiography.* Directory of Mark Twain's maxims, quotations, and various opinions. http://www.twainquotes.com/Lies.html. Accessed 4 Apr 2016.

Tweed, David. 2015. What do weak oil prices mean for the South China Sea? Bloomberg.com, 20 Jan 2015. http://www.bloomberg.com/news/articles/2015-01-20/all-about-the-base-oil. Accessed 17 May 2015.

United States Census Bureau (U.S. Census Bureau). 2015. Foreign trade – U.S. top trading partners. *Census.Gov.* https://www.census.gov/foreign-trade/statistics/highlights/top/top1512yr.html. Accessed 20 Mar 2016.

United States Department of Treasury (U.S. Treasury). 2016. Treasury International Capital (TIC) System – Home Page. https://www.treasury.gov/resource-center/data-chart-center/tic/Pages/index.aspx. Accessed 20 Mar 2016.

United States Energy Information Administration (US EIA). 2013. South China Sea. http://www.eia.gov/beta/international/regions-topics.cfm?RegionTopicID=SCS. Accessed 23 Jul 2015.

United States, Office of the Press Secretary, the White House (U.S. Press Secretary). 2016. Remarks by President Obama At U.S.-ASEAN press conference, 16 Feb 2016. Home, Briefing room, speeches & remarks. https://www.whitehouse.gov/the-press-office/2016/02/16/remarks-president-obama-us-asean-press-conference. Accessed 20 Mar 2016.

Vietnam National Administration of Tourism (Viet Nam Tourism). 2015. International visitors to Viet Nam in March and 3 months of 2015, 27 Mar 2015. http://vietnamtourism.gov.vn/english/index.php/items/8521. Accessed 7 Apr 2016.

Vu Trong Khanh, and Jenny W. Hsu. 2014. Anger over oil rig turns violent In Vietnam. *The Wall Street Journal*, 14 May 2014. http://www.wsj.com/articles/SB10001424052702303851804579560923647957180. Accessed 17 Mar 2016.

Webster, Thomas J. 2009. *Introduction to game theory in business and economics.* Armonk: M.E. Sharpe.

Whaley, F. 2016. *U.S. and Philippines bolster air and sea patrols in South China Sea.* New York Times, 14 Apr 2016. http://www.nytimes.com/2016/04/15/world/asia/south-china-sea-philippines-us-naval-patrols.html?_r=0. Accessed 19 Apr 2016.

Wiemer, Calla. 2009. Economic corridors for the Greater Mekong Subregion. East Asian Institute, Singapore, EAI background brief No. 479, 15 Sept 2009. http://www.eai.nus.edu.sg/publications/files/BB479.pdf. Accessed 6 Apr 2016.

Wilson, Brian, and Roger Christian. 1964. Don't worry baby. Attribution available. http://www.rollingstone.com/music/lists/the-500-greatest-songs-of-all-time-20110407/the-beach-boys-dont-worry-baby-20110526; lyrics available http://beachboyssongs.com/lyrics-page-1/lyrics-dont-worry-baby/. Accessed 6 Mar 2016.

Workman, Daniel. 2016. China's top 10 exports. World's top exports, 22 Mar 2016. http://www.worldstopexports.com/chinas-top-10-exports/1952. Accessed 7 Apr 2016.

World Bank, World Development Indicators. http://databank.worldbank.org/data/views/reports/metadataview.aspx

World Bank. 2010. Foreign direct investment – The China story. The World Bank, 16 Jul 2010. http://www.worldbank.org/en/news/feature/2010/07/16/foreign-direct-investment-china-story. Accessed 18 Apr 2015.

World Trade Organization (WTO)-Institute of Developing Economies (IDE-JETRO). 2011. Trade patterns and global value chains in East Asia: From trade in goods to trade in tasks. https://www.wto.org/english/res_e/booksp_e/stat_tradepat_globvalchains_e.pdf. Accessed 8 Feb 2015.

Xiaokun, Li. 2013. China sails through 'first island chain'. 2013-08-02 China Daily USA. http://usa.chinadaily.com.cn/china/2013-08/02/content_16863855.htm. Accessed 16 Nov 2014.

Xu, Beina. 2014. South China Sea tensions. Council on Foreign Relations—CFR Backgrounders, 14 May 2014. http://www.cfr.org/china/south-china-sea-tensions/p29790. Accessed 22 May 2015.

Yam, Tom. 2012. Game theory and the South China Sea conflict. South China Morning Post, 16 Nov 2014. http://www.scmp.com/article/1006824/game-theory-and-south-china-sea-conflict. Accessed 16 Nov 2014.

Zheng, Shibo. 2014. Interview: ASEAN welcomes China's new Maritime Silk Road initiatives—Guangxi Beibu Gulf Economic Zone. Xinhuanet, 15 Aug 2014. http://news.xinhuanet.com/english/china/2014-08/15/c_133559198.htm. Accessed 2 Oct 2014.

Zhong, Xinhui. 2013. The gaming among China, the Philippines and the US in the South China Sea disputes. Master Thesis, Development and International Relations, Aalborg University, Denmark. June 2013. http://projekter.aau.dk/projekter/files/76994735/Thesis_final_edition.pdf. Accessed 16 Nov 2014.

Index

D.J. Green, *The Third Option for the South China Sea*,
DOI 10.1007/978-3-319-40274-1

The manufacturer's authorised representative in the EU is Springer Nature Customer Service Centre GmbH, Europaplatz 3, 69115 Heidelberg, Germany. If you have any concerns regarding our products, please contact ProductSafety@springernature.com

Printed and bound by CPI Group (UK) Ltd, Croydon, CR0 4YY
30/09/2026
02225666-0013